Voichita Bucur

Urban Forest Acoustics

Voichita Bucur

Urban Forest Acoustics

With 109 Figures and 33 Tables

Springer

Voichita Bucur
Chargée de Recherches à l'INRA
Professeur honoraire
INRA – Centre de Nancy
LERMAB – Laboratoire d'Études et de Recherches
sur le Matériau Bois
UMR INRA 1093
54280 Champenoux
France

Library of Congress Control Number: 2005938507

ISBN-10 3-540-30783-4 Springer Berlin Heidelberg New York
ISBN-13 978-3-540-30783-9 Springer Berlin Heidelberg New York

Springer is a part of Springer Science + Business Media
springer.com

Editor: Dr. Dieter Czeschlik, Heidelberg, Germany
Desk editor: Dr. Andrea Schlitzberger, Heidelberg, Germany
Cover design: *design & production*, Heidelberg, Germany
Typesetting and production: LE-TeX Jelonek, Schmidt & Vöckler GbR, Leipzig, Germany
31/3100/YL – 5 4 3 2 1 0 – Printed on acid-free paper

Preface

In general, trees are viewed as admired symbolic individuals, producing recreational, spiritual and emotional rejuvenation. Their lifespan can far exceed that of humans. Planting a tree is a singular act of faith in the future, creating a legacy for the community members who will follow. The presence of trees in an urban area has been a reality for several centuries. Beautiful trees in urban plazas are synonymous with a high sense of community and civic pride. Trees significantly enhance the landscaping and appearance of the built environment.

City trees improve several architectural and engineering functions, providing a green infrastructure for communities. Trees create a friendlier environment for walking, riding bikes and working, by reducing glare and softening harsh traffic sounds and concrete views. Trees enhance the viewing in urban areas of a variety of birds and small animals, such as squirrels. They are of extreme importance to the functioning of many different ecosystems. Trees planted in the right place around buildings can improve air conditioning and heating costs by providing shade or by affecting wind speed or direction. Evergreen trees with dense, persistent needles can be used to provide a windbreak, while deciduous trees allow the sun to warm a house in winter. The more compact the branches and foliage of a group of trees, the greater their influence as a windbreak. It has been shown that trees are able to remove pollutants from the air; and they are seen as an important potential resource for removing greenhouse gasses from the atmosphere. Trees contribute to the protection of the environment and public health, providing economic and social benefits, encouraging positive social interaction.

In a modern concept, urban forest refers to all trees and vegetation in urban and suburban areas.

My motivation for writing this book comes from the frequently asked questions about urban environmental integrity, related namely to noise, climate, air and water quality.

This book is structured in nine chapters. As usual the first chapter "Introduction" relates the concept of the urban tree in contrast to the forest tree and gives a short description of the dendrological characteristics of different trees in the urban environment. The second chapter is "Noise in Forest" and refers to sound propagation in forest and the factors affecting this propagation. The equipment for in situ noise measurement is presented. The third chapter introduces acoustical sensors for the measurement of tree characteristics (diameter, height, mechanical and genetic characteristics). Chapter 4 is devoted to noise attenuation with plants, setting aside ground attenuation, scattering by trees, foliage, trunks and branches. The last section of this chapter refers to reverberation and attenuation in a forest stand. Chapter 5 depicts a very current subject, namely, protection against traffic noise from highways, railways and aircraft. Chapter 6 – noise abatement and dwellings in urban and suburban areas – underlines the necessity to take into consideration the meanings of the soundscape, which are environmental, historical or cultural. The practical application of this concept produces sound maps for urban planning. A positive impression on the urban soundscape is produced by large vegetation areas, belts of trees, public gardens and parks. Chapter 7 offers a brief discussion on the relationships between noise, animals, insects and trees and, of course, the acoustic methods for the detection of the presence of these biological agents in different stages of development. Chapter 8 – fire control with acoustical methods – briefly describes the potential of acoustics in forest fire detection and control. Finally, it seems appropriate to end this book (Chap. 9) with some considerations about economic aspects related to the value of urban trees.

Acknowledgements

First of all I wish to acknowledge the National Institute for Agricultural Research (INRA France) Forestry Research Center in Nancy–Champenoux, and the University Henri Poincaré, Nancy 1, Faculty of Science, Wood Research Laboratory, for providing facilities for writing this book.

I am indebted to different organizations and individuals cited in this book for permission to reproduce figures and tables. In my bibliographic research, I was assisted by the kind collaboration of helpers at various libraries in France, such as the library of the University "Henri Poincaré" in Nancy – Marie Annick Bruthiaux, the library of "Ecole Nationale des Eaux et Forêts de Nancy – Marie Jeanne Lionnet, David Gasparotto and Bruno Spandonide, while he was a student at the University Paris Sorbonne. I am very much indebted to Professor Helmut Resch and to David Gasparotto for comments that improved the final manuscript of this book. Also I wish to acknowledge my colleagues Dr Laurent Chrusciel, Dr Stéphane Dumarcay, the PhD students Youcef Irmouli and Anthony Dufour and our secretaries Corinne Courtehoux and Catherine Antoni, for everyday assistance with the electronic form of the manuscript. Thanks are due to my sister Despina Spandonide for continuous and enthusiastic encouragement during the writing this book. Last but not least I wish to acknowledge Constantin Spandonide for his generosity, spending many hours in preparing the figures for this book.

Finally, I wish to express my sincere gratitude and admiration to the staff of Springer Verlag while working with me for the final version of the manuscript of this book, for the pleasant and enjoyable professional moments we spent together via the modern communication media.

Champenoux, January 2006 *Voichita Bucur*

Contents

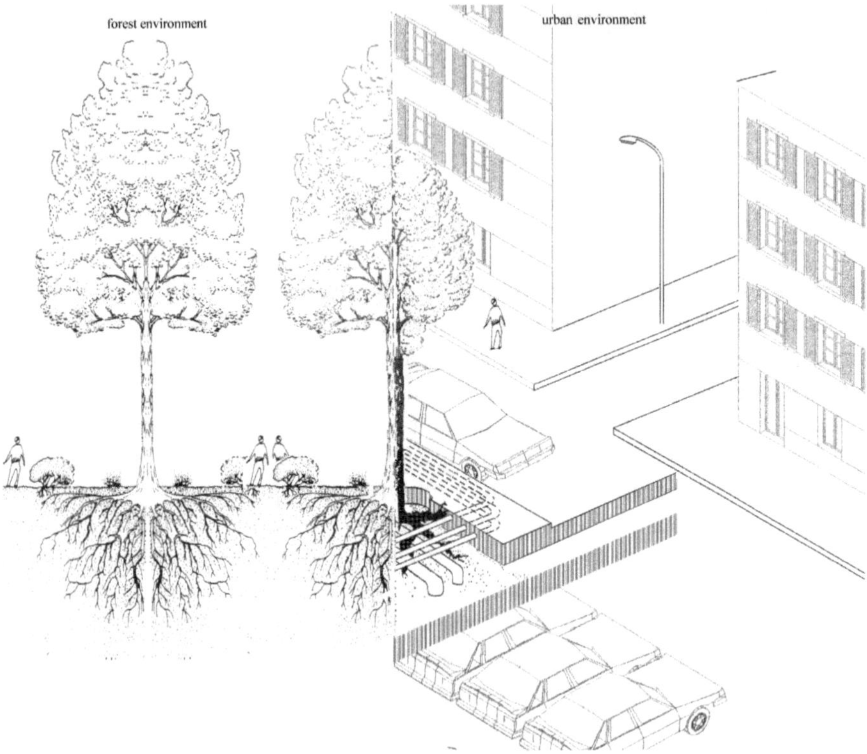

Fig. 1.1. Trees in natural and urban environments

name (species), e.g.: *Abies alba*. The specific name can be traced to several origins: Latin, Celtic, Greek, etc. (so *alba* from Latin = white; Aymonin 1986). The common name for *Abies alba* is fir, which is generally used and has been handed down from generation to generation. In this book, both scientific and common names will be used.

Considerable ecological and silvicultural information has been developed in reference books and manuals in the past century for judging how a tree or a stand should be managed. Specific features for individual trees and stands must be considered. The main criteria to select trees for urban and suburban areas are related to the growth and silvicultural requirements of each species. Following the position of a tree in a stand, trees can be classified as:

- dominant trees, with well formed crowns, receiving sunlight uniformly
- co-dominant trees, in the high canopy
- intermediate trees with crowns in the lower part of the canopy, shaded by the surroundings
- suppressed trees, with crowns below the main level of the canopy.

Fig. 1.2. Silhouettes of trees (from Hosie 1969; reproduced by permission of Natural Resources Canada, Canadian Forest Service, copyright 2005)

For each tree, the morphological and physical characteristics which must be considered are the following: height, diameter at breast height, growth ratio (radial increment rate), live crown ratio (height of crown divided by total tree height), density expressed as number of stems per hectare and general health aspect – the vigorous aspect of the tree, without insect damage or disease.

Remarkable studies by Zimmermann and Brown (1980), Wilson (1984) and Mattheck (1996, 1998) enable the reader to increase his questions and answers related to the biological and mechanical functions of trees.

Identification of native trees and plants is essential for the prediction of better growth conditions of trees in a specific site. The introduction of species like ornamental apples and cherries is used to develop the existing vegetation quickly and to satisfy the socio-economic requirements of the residents.

The street tree population is very variable and is composed of hardwoods and mixed softwoods/hardwoods, having a density of 100 trees/km of street and a diameter ranging from 10 cm to 60 cm. Deciduous trees ensure greater water evaporation and consequent cooling of the street, while mixed trees ensure a higher noise attenuation efficiency because of the evergreen species used. The diversity of urban morphology determines the structure of street tree patterns, related to the natural environment and the management policies of cities and adjacent residential or suburban zones. Table 1.1 gives some dendrometric characteristics of different species from the temperate zone.

Rapid urbanization after the First and Second World Wars altered the microclimate in urban areas, through a gradual replacement of original forest by man-made buildings and structures which increased the heat-storage capacity of cities. Street trees, as well as parks, gardens and green spaces, are natural air

Table 1.1. Some dendrological characteristics of several species growing in a forest environment (data from Hora 1981; Aymonin 1986)

Species		Height	Age (years)	
Scientific name	Common name	(m)	Maturity	Longevity
	Deciduous species			
Acer pseudoplatanus	Sycamore	30	25	200–500
Aesculus hypocastanum	Horse chestnut	25	24	200
Fagus silvatica	Common beech	45	30	300
Liriodendron tulipifera	Tulip tree	60	30	500
Quercus robur	Oak	25	45	2,000
Betula pendula	Birch	15	10	100
Populus alba	White poplar	10	5	50
Tillia cordata	Lime	35	20	500
	Coniferous species			
Picea abies	Spruce	50	50	400
Abies alba	Fir	50	15	200
Pinus strobus	Eastern white pine	80	50	200
Pinus contorta	Lodgepole pine	33	Unknown	150
Thuja plicata	Arbor vitae	60	Unknown	400
Larix decidua	Larch	35	Unknown	600
Chamaecyparis lawsoniana	False cypress	50	Unknown	400
Sequoia sempervirens	Redwood	120	Unknown	2,000

Table 1.2. Noise reduction with different patterns of street trees in Nanjing, China (data from Mao et al. 1993). The tree species are: *P.a.* = *Platanus acerifolia*; *M.g.* = *Metasequoia glyptostroboides*; *S.c.* = *Sabina chinesis*; *P.t.* = *Pittosporum tobira*; *C.i.* = *Carya illinoensis*; *C.d.* = *Cedrus deodara*; *E.j.* = *Euonymus japonica*

Parameters	Streets			
	No 1	No 2	No 3	No 4
Street width (m)	40	42	28	30
Tree pattern	Deciduous	Mixed	Deciduous	Mixed
Number of tree rows	6	4	2	4
Width of green belt (m)	35	35	2	4
Canopy height (m)	4–25	4–22	4–25	4–20
Crown projection (%)	80–85	80–85	85–90	80–85
Tree species	*P.a*	*M.g.; S.c.; P.t.; C.i.; E.j.*	*P.a.*	*M.g.; C.d.; C.i*
Noise attenuation (dB)	6	4	1	8
Efficiency (dB/m)	0.24	0.31	0.10	0.36

conditioners and, within a limited range, noise attenuators. Mecklenberk et al. (1972) noted that the noise attenuation capacity of trees is directly related to the density and width of planting zones. The efficiency of noise attenuation, as expressed in Table 1.2, is 0.36 dB/m for mixed zones and only 0.17 dB/m for zones planted with only one species.

The existing information in the literature on noise reduction in urban environment is quite abundantly disseminated in publications related to forest and agricultural studies during the period 1970–1990 and is very scarce later; and, in contrast, publications related to acoustic studies during the past 20 years stress the development of modeling techniques. The aim of this book is to show the necessity of understanding both aspects.

2 Noise in Urban Forest

Noise in urban forest is produced by the sound field of different sources which can be detected in the surroundings. The acoustic intensity of this field is characterized by the following parameters: the amplitude of the disturbance, the excess pressure, the particle velocity, the density change or corresponding change in refractive index, the steady pressure on a surface due to the impact of sound waves, the thermal changes produced by alternating compression and rarefaction and the power which may be absorbed from the sound waves. From a theoretical point of view, three fundamental types of sources are recognized: the simple point source, the doublet (or dipole, equivalent to two simple and equal sound sources), and the quadrupole (the combination of two doublet sources, termed longitudinal and lateral quadrupole). A simple point sound source can be produced by a single-shot propane gun source.

To study impulse source scattering in forest, Rogers et al. (1992) used a propane gun, which contains a significant amount of low-frequency acoustic energy, and a microphone located in a stand, at 10 m from the source. It was observed that the received signal is composed of two main components: (a) a direct zone produced by sound wave direct propagation from the source to the microphone and (b) a scattered zone induced by the presence of woods. Scattering phenomena in a stand are very complex and rather difficult to estimate accurately. In order to make a detailed assessment of the influences of all factors producing scattering in a forest stand (biomass, density of trees/ha, tree height, tree diameter, crown shape and size, size and shape of leaves and needles, etc.), it has been accepted to study a global parameter expressed by the excess attenuation, which includes the absorption, dispersion, reflection and refraction of sound.

The specification of noise in physical terms depends upon its nature. One of the best representations is given by its spectrum. For noise measurement, three techniques are used: recording the wave-form to identify the disturbing frequency components, narrowband analysis and broadband analysis when determining the requirements for noise control. For most purposes, it is sufficiently accurate to use octave band analysis.

In the first part of this chapter, several acoustical notions necessary for the understanding of the theoretical and practical approaches are proposed. Factors affecting sound propagation and scattering phenomena are discussed.

The second part of this section is devoted to a presentation of the equipment for noise measurements.

2.1 Sound Propagation

2.1.1 Definitions and Theoretical Considerations

The sound is produced by a disturbance induced in air, causing alternative pressure and displacement of the air molecules. The dictionary of acoustics (Morfey 2001) and basic reference books (Stephens and Bate 1966; Beranek and Vèr 1992; Fahy and Walker 1998; Harris 1998; see also sources for noise level data in journals such as: Acta Acustica, J Acoust Soc Am, J Sound Vibr, Noise Control Eng J; and the US National Bureau of Standards and the ISO standards noted in Annex 4), in an acoustical context, define noise as an undesired and extraneous sound. A sound wave can be composed of a single frequency (pure tone), or a combination of this frequency harmonically related or not.

The measurable aspects of sound propagation in air can be described by many parameters. In this book, I selected only 12 parameters, as follows:

1. *Sound pressure* is the variation in pressure above and below atmospheric pressure and is expressed in Pascals (Pa). The normal audible frequency range is roughly between 15 Hz and 16 kHz. Frequencies between 3 kHz and 6 kHz are the most sensitive. A young person can detect pressure as low as 20 μPa, compared to normal atmospheric pressure, which is 101.3×10^3 Pa.
2. *Speed of sound* in air (noted c in m/s) is calculated as:

$$c = \sqrt{\frac{1.4P_s}{\varrho}} \tag{2.1}$$

   where P_s is the ambient pressure (Pa) and ϱ is the air density (kg/m^3). The speed of sound in air is dependent on temperature. Some theoretical aspects related to this interaction are presented in Annex 3.
   For practical purposes, the speed of sound is determined with the following approximate formula:

$$c = 331.4 + 0.607\theta \tag{2.2}$$

   where θ is the ambient temperature in °C, or with the exact formula:

$$c = 331.4\sqrt{\frac{T}{273}} = 331.4 + \sqrt{1 + \frac{\theta}{273}} \tag{2.3}$$

where T is the absolute temperature (K). At the normal temperature of 20 °C, the speed of sound is 344.8 m/s.

3. *Sound intensity* (W/m^2) is the sound energy transmitted through a specific area and measured in a specific direction. In free space, the sound intensity is related to the total power radiated into the air by a sound source and to the sound pressure. Sound intensity at a point is a vector, having a minimum and a maximum. The maximum is obtained when its plane is perpendicular to the direction of travel; when parallel, the sound intensity is zero. The sound intensity is related to the sound pressure. In an environment without reflecting surfaces, at any point, the sound pressure of freely traveling waves (plane, cylindrical, spherical) is related to the maximum intensity I_{max}, through the equation:

$$I_{max} = \frac{p_{rms}^2}{\varrho \cdot c} \tag{2.4}$$

where p_{rms} is the root-mean-square (rms) sound pressure (expressed in Pa or N/m^2), ϱ is the density of the air (kg/m^3), c is the speed of sound in air (m/s), $c\varrho$ is the characteristic impedance of the air $\left(\frac{m}{s} \cdot \frac{kg}{m^3}\right)$.

4. *Sound power level* is the measure of the total acoustic power radiated by a source and is expressed in dB *re* W_0, which is the reference sound power, standardized at 10^{-12} W, and is defined as:

$$L_W = 10 \log_{10} W/W_0 (\text{dB}\, re\, W_0) \tag{2.5}$$

where W is the sound power (W) and W_0 is the reference sound power, standardized at 10^{-12} W, corresponding to the reference pressure of 20 µPa (2×10^{-5} N/m^2).

The relationships between the sound power and sound power level are given in Table 2.1, from which it can be seen that power ratio < 1 lead to negative levels. Different international standards describe methods for determining the sound power levels of noise sources (see Annex 4).

5. *Sound intensity level* noted IL or L_I (dB) is the measure of the acoustical disturbance produced at a point removed from the source and is defined as the ratio of two sound sources intensities, I_1 and $I_2 = I_{ref}$ expressed in logarithmic form as:

$$IL = L_I = 10 \log_{10} \frac{I_1}{I_{ref}} \tag{2.6}$$

where I_{ref} is the reference intensity of 10^{-12} W/m^2 (if the reference is different, one must note explicitly the reference value). The sound intensity level depends on the distance from the source and the losses in the air path (ISO 3740, ISO 3744; see Annex 4).

Table 2.1. Sound power level (dB) and sound radiated power (W) in linear, exponential and dB-log scale (data from Beranek 1960, 1992)

Sound radiated power (W)		Sound power level (L_W, dB)	
Usual notation	Exponential notation	Relative to 1 W	Relative to 10^{-12} W
100 000	10^5	50	170
1,000	10^3	30	150
100	10^2	20	140
10	10^1	10	130
1	1	0	120
0.1	10^{-1}	−10	110
0.01	10^{-2}	−20	100
0.001	10^{-3}	−30	90
0.00001	10^{-5}	−50	70

6. *Sound pressure level* noted SPL or L_p (dB) is the ratio between the effective measured sound pressure and the sound pressure at a source reference:

$$L_p = 10 \log_{10} \frac{\overline{p^2(t)}}{p^2_{ref}} = 20 \log_{10} \frac{p_{rms}}{p_{reference}} \tag{2.7}$$

where $p_{reference}$ is the reference pressure of 20 µPa (2×10^{-5} N/m^2), for sound propagation in air, since it corresponds to the rms pressure of a pure tone at 1 kHz, which is just audible by the human ear. The rms corresponds to the acoustic pressure fluctuations of the acoustic wave and is given by the equation:

$$\overline{p^2(t)} = \underline{\lim_{T \to \infty} \frac{1}{T} \int_{-T/2}^{T/2} p^2(t)\,dt} \tag{2.8}$$

where T is the averaging time, very large compared to the period of pressure fluctuation and should extend to infinity for random fluctuations, whose statistical properties remain stationary with time. Since this parameter has the dimensions of pressure squared, the label "root mean square" was associated with this fluctuation. The parameter p_{rms} is given by the square root of the mean square pressure. In practice the range of variation of p_{rms} is very large, from 10^{-5} Pa to 10^3 Pa. For this reason, the logarithmic scale is always used.
Typical values of the rms pressure fluctuation and the corresponding sound pressure levels are given in Table 2.2.
The sound pressure level at different frequencies produced by different sources (wind, cars, train, etc.) is given in Table 2.3.

Table 2.2. Typical rms pressure fluctuations and their sound pressure levels (Fahy and Walker 1998, with permission)

Source	Pressure fluctuation p_{rms} (Pa)	Sound pressure level L_p (dB *re* 2×10^{-5} Pa)
Jet engine at 3 m	200	140
Pneumatic hammer at 2 m	2	100
Conversational speech	0.02	60
Residential area at night	0.002	40
Rustling of leaves	0.0002	20
Threshold of hearing	0.00002	0

Table 2.3. Noise data at octave-band center frequency for different noise sources (Egan 1988)

Source	Sound pressure level (dB) at various frequencies (Hz)								SPL
	63	125	250	500	1,000	2,000	4,000	8,000	dB
Birds at 33 m	–	–	–	–	–	50	52	54	57
Cicadas	–	–	–	–	35	51	54	48	57
Large dog at 17 m	–	50	58	68	70	64	52	48	72
Lawn mower at 1.7 m	85	87	86	84	81	74	70	72	86
Pistol shot at 82 m	–	–	–	83	91	99	102	106	106
Surf at 3 m, moderate sea	71	72	70	71	67	64	58	54	78
Wind in trees, 16 km/h	–	–	–	33	35	37	37	35	43
Large trucks	83	85	83	85	81	76	72	65	86
Passenger cars	72	70	67	66	67	66	59	54	71
Motorcycle	95	95	91	91	91	87	87	85	95
Snowmobile	65	82	84	75	78	77	79	69	85
Train at 33 m	95	102	94	90	86	87	83	79	94
Car horn at 5 m	–	–	–	92	95	90	80	60	97
Commercial turbofan airplane	77	82	82	78	70	56	–	–	79
Military helicopter	92	89	83	81	76	72	62	51	80

The relation between sound pressure in microPascals and sound pressure level in decibels (re 20 μPa) for various sources of noise is given in Fig. 2.1. All confusion between sound power level (often expressed in Bels) and sound pressure level (expressed in dB) must be avoided. The former corresponds to the measure of the acoustic power radiated by the source and the later depends on the power of the source, the distance from the source and the acoustical characteristics of the space surrounding the source.

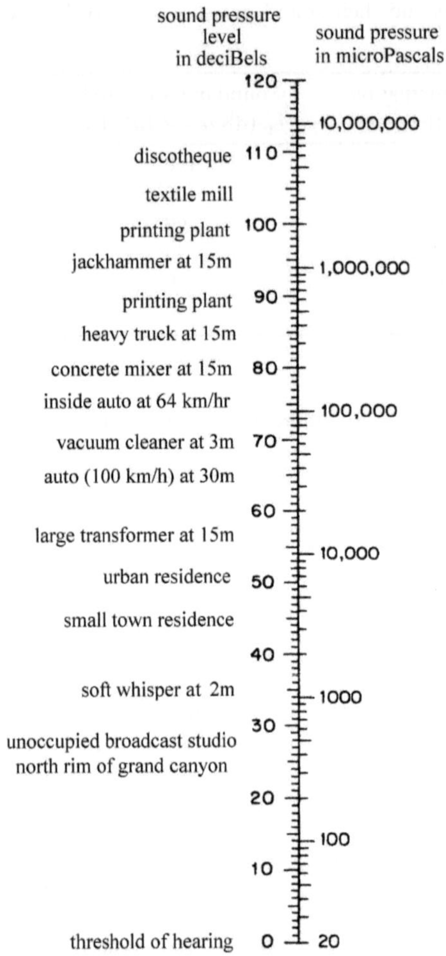

Fig. 2.1. Various sources of noise and the corresponding sound pressure level (dB) and sound pressure (μPa; Harris 1998). Reprinted with permission from the Acoustical Society of America, copyright 2005

7. *A-weighted sound pressure level* (L_A, in dB) is defined as:

$$L_A = 10 \log \left[\frac{p_A(t)}{p_{ref}} \right] \tag{2.9}$$

where $p_A(t)$ is the instantaneous sound pressure measured using the standard frequency-weighting A.
8. *Average sound level* ($L_{av,T}$) during time T, is expressed in decibels and is defined as:

$$L_{\mathrm{av},T} = 10\log\frac{\frac{1}{T}\int\limits_0^T p^2(t)\,\mathrm{d}t}{p_{\mathrm{ref}}^2} \tag{2.10}$$

where T is the long time over which the averaging takes place (e.g. 8 h).

9. *Averaged A-weighted sound level* ($L_{\mathrm{eq}} = L_{\mathrm{A},T}$) is defined as:

$$L_{\mathrm{eq}} = L_{\mathrm{A},T} = 10\log\frac{\frac{1}{T}\int\limits_0^T p_{\mathrm{A}}^2(t)\,\mathrm{d}t}{p_{\mathrm{ref}}^2} \tag{2.11}$$

where T is 8 h for a working day or 24 h for a full day.

10. *Day night noise level* ($L_{\mathrm{d,n}}$), between 07.00 and 22.00 hours (dB) is given by:

$$L_{\mathrm{d,n}} = 10\log\frac{1}{24}\left[\frac{\int\limits_{7:00}^{22:00} p_{\mathrm{A}}^2(t)\,\mathrm{d}t}{p_{\mathrm{ref}}^2} + \frac{\int\limits_{22:00}^{7:00} 10p_{\mathrm{A}}^2\,\mathrm{d}t}{p_{\mathrm{ref}}^2}\right] \tag{2.12}$$

11. *A-weighted sound exposure* ($E_{\mathrm{A}T}$; Pa^2 s) is proportional to the energy flow (intensity × time) in a sound wave in the time-period between t_1 and t_2 and is given by:

$$E_{\mathrm{A},T} = \int\limits_{t_1}^{t_2} p_{\mathrm{A}}^2(t)\,\mathrm{d}t \tag{2.13}$$

12. *A-weighted noise exposure level* ($L_{\mathrm{EA},T}$) which is:

$$L_{\mathrm{EA},T} = 10\log\left(\frac{E_{\mathrm{A},T}}{E_0}\right) \tag{2.14}$$

where the reference E_0 at $(20\,\mu\mathrm{Pa})^2$ is equal to $(4\times10^{-10}\,\mathrm{Pa})^2\,\mathrm{s}$ (Beranek 1992) or, following ISO 1996-1 (see Annex 4), E_0 is equal to $(1.15\times10^{-10}\,\mathrm{Pa})^2\,\mathrm{s}$.

The system used for noise control contains three major components: the source, the path and the receiver, associated with emission, transmission and immission. The sound energy emitted by a noise source is transmitted to the receiver where it is immitted. The immission is described by the sound pressure level (dB). The strength of the noise source is described by the sound power level and its directivity, which is a function of angular position around source and frequency.

To match the assumed frequency response of the ear, implied by equal loudness contours, A, B and C frequency weighting curves were standardized.

Table 2.4. Electrical weighting networks for sound-level meter, for several frequencies (Beranek 1992)

		Frequency (Hz)								
		10	20	50	100	200	400	1,000	10,000	20,000
A-weighting	dB	−70.4	−50.5	−30.2	−19.1	−10.9	−4.8	0	−2.5	−9.3
C-weighting	dB	−14.3	−6.2	−1.3	−0.3	0	0	0	−4.4	−11.2

The network specification of these curves is given in Annex 4. The A-frequency weighting was for 40 dB sound level, the B-frequency weighting for 70 dB sound level and C-frequency weighting for 100 dB level. For outdoor community noise measurements, A-frequency weighting is mostly used. This weighting reduces the sensitivity of the sound level meter to low and high frequency sounds, as compared with the mid-band frequency, between 1 kHz and 4 kHz. Another important practical advantage of A-weighting is the relative immunity against wind noise generated at the microphone (Fahy and Walker 1998). Today, the B and C frequency weightings are out of use. The time of weighting can be fast (F), 125 ms (corresponding approximately to the ear integration time), slow (S), with an exponential time constant of 1 s, and impulse (I), which has a fast rise (35 ms) and slow decay.

The A-weighted sound or noise power level is defined as:

$$L_{WA} = \log \frac{W_A}{W_0} \tag{2.15}$$

where W_A is the A-weighted sound power and W_0 is the reference sound power, 10^{-12} Watt. Table 2.4 gives the A-weighted sound pressure level for different frequencies. When A-weighting is used with the overall sound power level, the noise power emission level can be expressed in Bells or decibels (1 Bel = 10 dB).

The radiation source field varies with distance from source, which can be in the near field, far field or reverberant field. In the far field, the sound pressure decreases by 6 dB for each doubling of the distance from the source. In outdoor measurements, the sound power must be performed in the far field of the source (choosing an array of microphone positions over the surface).

2.1.2 Factors Affecting Sound Propagation

Sound or noise propagation in a forest stand is affected by the presence of trees, soil surface, ground vegetation, topography and meteorology of the site. The interaction between the sound field and the vegetation is complex and determines mainly the decrease in the sound level, but sometimes can induce a small increase in the sound level under the canopy. The presence of a solid barrier between the source and receiver introduces a much more complex mechanism of sound propagation, producing a decrease in sound level behind the canopy and barrier with another possible zone of unstable increased or decreased sound level, depending on particular in situ conditions.

Acoustic scattering in woodlands was a subject of interest for many authors (Eyring 1946; Embelton 1963; Aylor 1972a, b; Decourt 1975; Fricke 1984; Price 1988; Attenborough 1992; Barrière and Gabillet 1999), with the principal focus on the attenuation of sound between the source and receiver induced by geometric spreading and attenuation effects due to absorption where the ground effect plays an important role. Scattering effects will be largely covered in Chap. 4.

To study the influence of meteorological conditions, Heimann (2003) proposed a simulation of wind propagation through an idealized stand of trees with a tri-dimensional numerical fluid-dynamics model which allowed isolation of single influences (trees, wind, ground, etc.) in a virtual way.

In atmosphere (Piercy et al. 1977; Brown 1987; Naz et al. 1992), sound undergoes the following physical processes: reflection from surfaces, refraction by temperature and wind gradients, diffraction by edges and changes in surface impedance, scattering by turbulent fluctuations in temperature, wind, rain and snow, molecular absorption and attenuation induced by scattering-out of finite angular-width beams. Relative humidity variations produce negligible changes (<3%) on the speed of sound. Near the ground, the propagation of sound waves (Noble et al. 1992) is very complex, involving: geometric spread and molecular absorption, reflection with phase change due to finite impedance of the ground, refraction by the mean wind and temperature gradient and scattering by atmospheric turbulence, at small and large scale.

Very roughly, the meteorological effects (Ingård 1953) can be summarized as follows: for a windy day and wind speed of 6–11 m/s, the variation in SPL was 15–20 dB at 2 kHz frequency; and the average attenuation was 4–6 dB over 100 m, with a maximum of 20 dB. Noble et al. (1992) noted that "large scale atmospheric features have a large effect on phase but little effect on the amplitude of the signal".

2.2 Equipment for Noise Measurement

2.2.1 Instrumentation and Noise Sources

The techniques and instruments for the measurement of surrounding noise are determinants for noise control and abatement at any point in the acoustic field as a function of time and frequency. At any observation point in the acoustic field, the microphone (which is an electroacoustic transducer) transforms sound pressure into a corresponding electrical signal. Environmental noise fluctuates greatly and it is essential to be able to measure acoustic phenomena accurately, with a good repeatability of readings with a sound-level meter. Sound-level meters are more or less complex, depending on their practical utilization which can be for:

- laboratory reference intended for calibration of other apparatus,
- precision sound level meters for laboratory and field accurate measurements,
- general purpose sound level meter for noise recording level and later data analysis, and survey sound level meter for noise environment.

The environmental acoustical signal is processed by the adapted devices which compose a sound-level meter. As presented in the literature and many commercial pamphlets (Fig. 2.2), the sound-level meter is a portable apparatus, battery-operated and equipped with a microphone, a pre-amplifier, an amplifier, a weighted network (A, B, C and linear) or an external bandpass filter, an output amplifier and a read-out meter, giving sound pressure level (L_p) and other parameters. Example: L_{10} = sound level exceeded 10% of the time; L_{50} = sound level exceeded 50% of the time; L_{90} = ambient level; L_A = sound level on A scale; L_{eq}, $L_{d,n}$ = day/night equivalent noise level. The microphone converts the incident acoustical signal into an electrical signal. Commonly, the dynamic range of sound pressure is 10^7 and for this reason a output voltage must also be provided. Most frequently, the attenuator is arranged in 10-dB steps. Frequency analysis of the sound field can be performed in one-third band analysis. Because of the very big fluctuations in sound level, the apparatus is provided with three responses, a "fast" response having a time constant of 10 ms, a "slow" response for 1 s and an "impulse" for a time constant of 35 ms. More complex data about the noise signal can be obtained when parameters such as the peak or duration of a transient (50 μs), the cross-power spectra, the computation of correlation functions, etc., are computed.

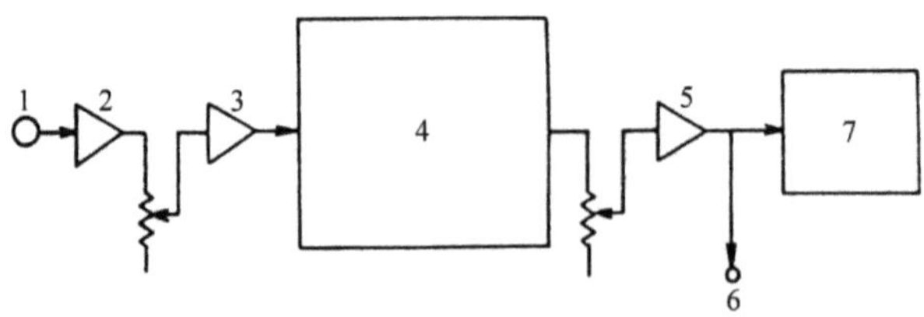

Fig. 2.2. Block diagram of a sound level meter. 1 Microphone, 2 pre-amplifier, 3 amplifier, 4 linear all-pass or weighted network (A, B, C) or external bandpass filter, 5 amplifier, 6 external output, 7 rectifier and read-out meter

It is very important to note that the sound-level meter must be checked regularly for acoustical calibration (accuracy 0.2 dB) and electrical calibration.

For scientific measurements and consistent data analysis, specific noise sources are used. Very popular is the single-shot propane gun source which contains a significant amount of low-frequency acoustic energy. A freon-powered horn was used by Rogers and Lee (1989) for in situ measurements. This source was located 60 m away from the forest and the microphone was located on a line from the source normal to the edge of the stand.

Also, as an acoustic source, a loudspeaker driven by a pre-recorded signal with traffic noise or white noise has been used (Fig. 2.3) for experimental measurements in a stand. For normal incidence, the source and the reference

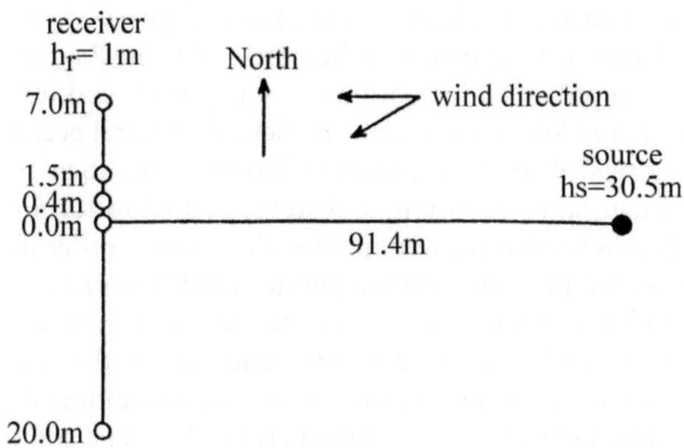

Fig. 2.3. Measurement geometry for in-field measurements on a flat surface (Noble et al. 1992). Reprinted with permission from the Acoustical Society of America, copyright 2005

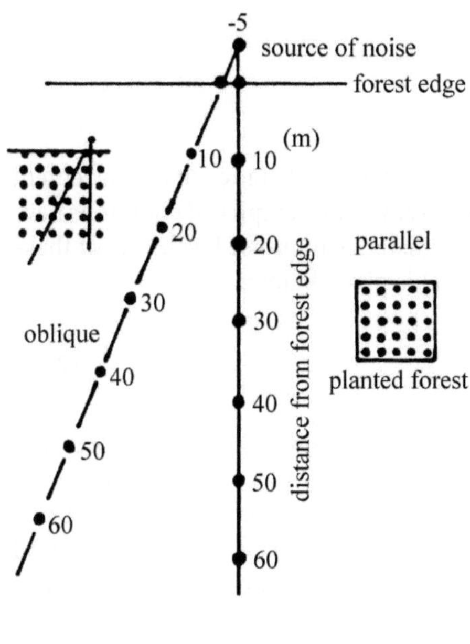

Fig. 2.4. Experimental arrangements for measurements in horizontal plane, in different stands (Tanaka et al. 1979)

microphone were positioned on a straight line (180°) at different distances from the ground (max 30.5 m). In order to control the experimental conditions, the sound level of the background noise must first be determined.

Measurements at an oblique incidence (Fig. 2.4) were performed by Tanaka et al. (1979) when the site configuration was very complex.

2.2.2 Measurement In Situ

Because of the diversity and complexity of experimental situations, in this section we propose to select only the aspects related to the amplitude and phase measured at the microphone, in situ, at 91 m from the source (Fig. 2.5).

The main difference between the amplitude variations and phase variations can be observed in a long-term excursion on the time-scale. The Fourier transform of phase variation represented in coordinates – amplitude and frequency – displays both short- and long-term variations. Several spectral peaks can be identified, probably related to the experimental conditions, such as length of sample analyzed and large atmospheric features related to turbulence. The peak at 0.0073 Hz shows that the dominant scale is on the order of hundreds of meters in size, corresponding to atmospheric or field features.

As a general remark, note that peaks must always be analyzed in terms of corresponding wave length, which can give an indication about the size and nature of the objects producing them. In practice, the understanding of those peaks requires fine equipment and very skilful operators having a good theoretical background.

2.2.2.1 Effects of Distance

The effects of scattering by woodlands related to the distance of measurement and frequency are addressed in the literature. Various empirical equations have been proposed to predict the influence of distance on noise level. One of these equations (Cook and van Haverbeke 1971) is given below:

$$S_d = S_0 - 20 \log \frac{d}{d_0} \tag{2.16}$$

where S_d is the sound level at distance d, S_0 is the sound level at the source and d_0 is the reference distance where the sound level is known. Figure 2.6 shows the influence of distance on decreasing sound level noise measurements in a tree belt (width of 30 m and in-row spacing of about 2 m). The belt was composed of deciduous trees of 25 m height. A very important decreasing effect was

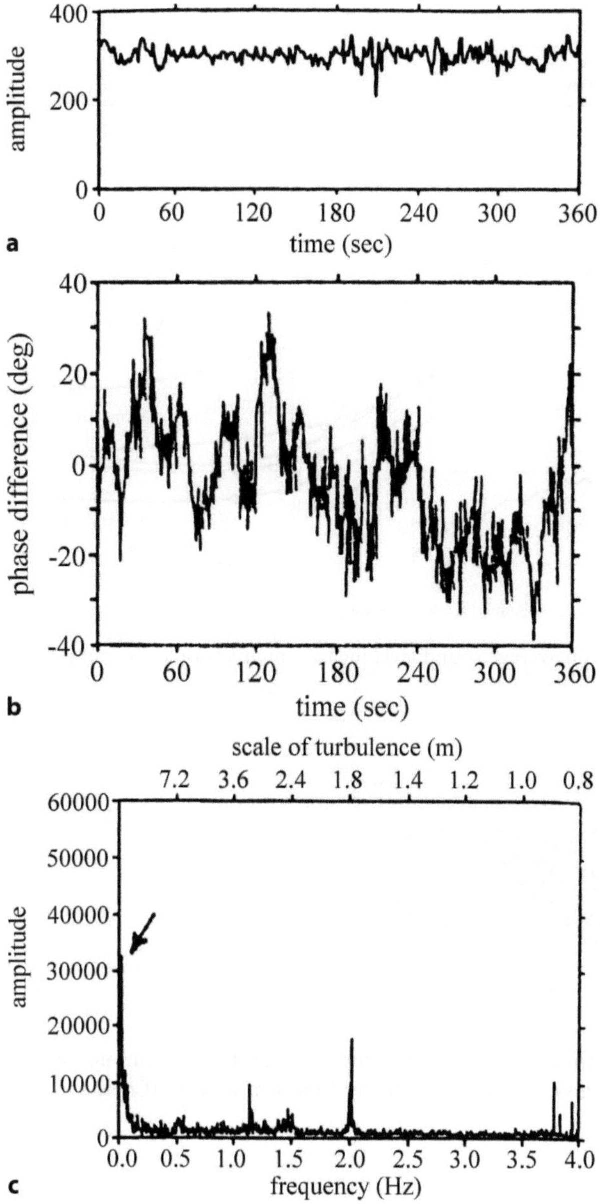

Fig. 2.5. Physical parameters measured at the microphone, situated 91 m from the source, on a grass field. The source is a loudspeaker (pre-registered traffic noise) located on top of a 30.5 m tower (Noble et al. 1992; reprinted with permission from the Acoustical Society of America, copyright 2005). **a** Amplitude versus time for 360 s, **b** phase difference vs time (360 s), at the receiver, **c** Fourier transform of the phase variation displayed in **b**, in the frequency range 0–4 Hz

observed for the distance between 33 m and 66 m, after which a quasi-constant level was observed.

To determine the optimum position of the tree belt as a noise screen, a series of experiments were performed, during which source and microphone position were varied simultaneously (Fig. 2.7). The vertical structure of the belt is

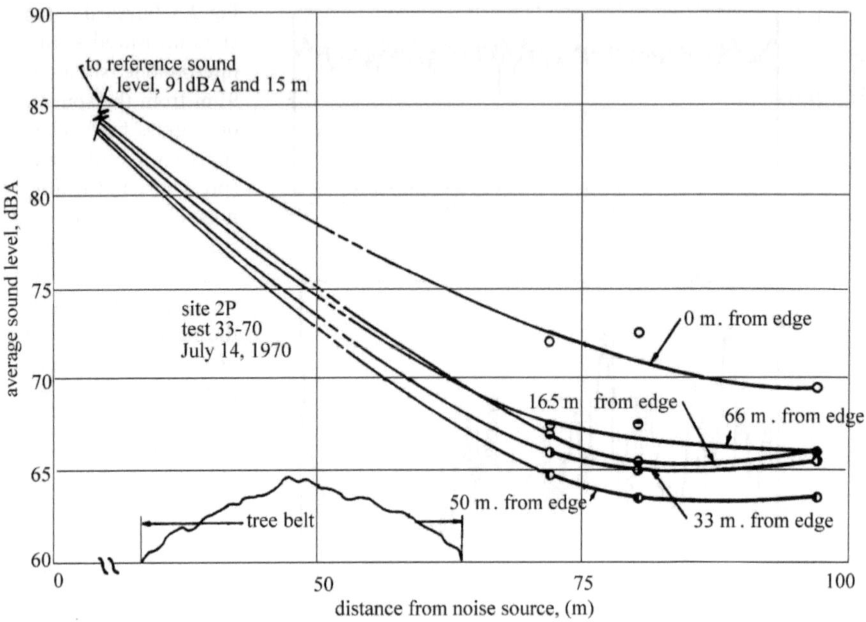

Fig. 2.6. Influence of distance on average sound level measurements (Cook and van Haverbeke 1971)

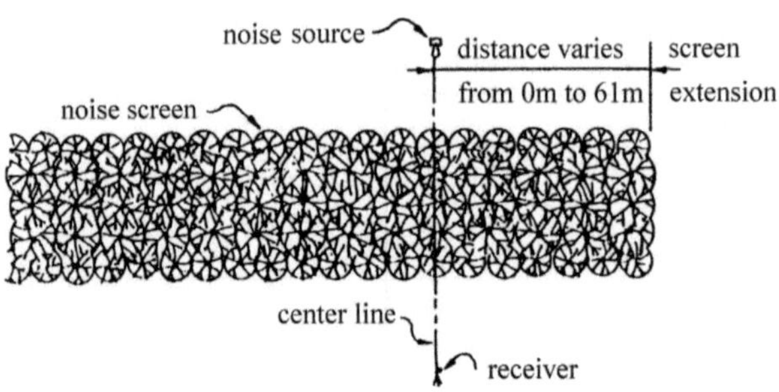

Fig. 2.7. Experimental arrangements for noise attenuation measurements through a tree belt, at different "screen extension" distances ranging from 0 m to 66 m (Cook and van Haverbeke 1971)

shown in Fig. 2.8. For quantifying the variation in relative attenuation, the reference variable was the ratio of the distance from receiver to source (R/S; Fig. 2.9). From this figure, one can see a minimum corresponding to R/S = 1, which corresponds to a tree belt situated midway between the source and the receiver. After this inflection point, the attenuation increases as the ratio R/S increases, indicating a more effective action of the tree belt. Cook and van Haverbeke (1971) noted: "it would seem that planting distances from 12 m

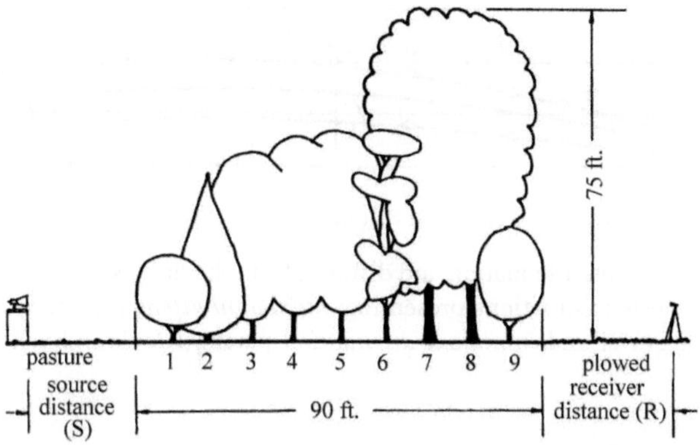

Fig. 2.8. Vertical structure of a belt of trees of width 30 m, of different species. Row spacing 3.3 m, in-row spacing 2.0 m (Cook and van Haverbeke 1971). 1 Russian olive, 2 pine and eastern red cedar, 3 catalpa, 4, 5 hackberry, 6 honey locust, 7, 8 cottonwood, 9 mulberry (90 ft = 27.45 m; 75 ft = 22.80 m)

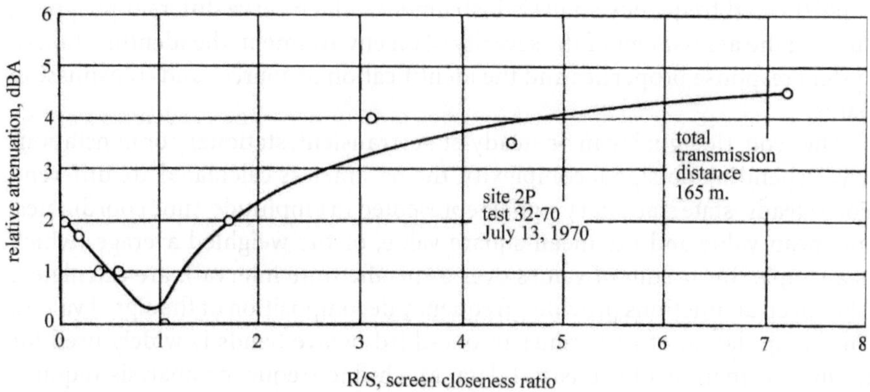

Fig. 2.9. Relative attenuation vs the ratio of receiver to source (R/S), for 165 m total transmission distance (Cook and van Haverbeke 1971)

to 22 m from the noise source would yield optimum results for tree belts of considerable height and depth in rural areas". Also: "placing trees and shrubs close to a noise source is recommended, a distance of 3 m to 8 m from a noise source to nearest shrub would seem to yield optimum results". The relative attenuation increases with the distance from the source (Fig. 2.10).

Beside the role of distance in outdoor sound transmission and therefore in noise reduction by tree belts, the roughness and acoustic impedance of the media interposed between the source and receiver play an important role (see also Chap. 4, the section related to ground effect).

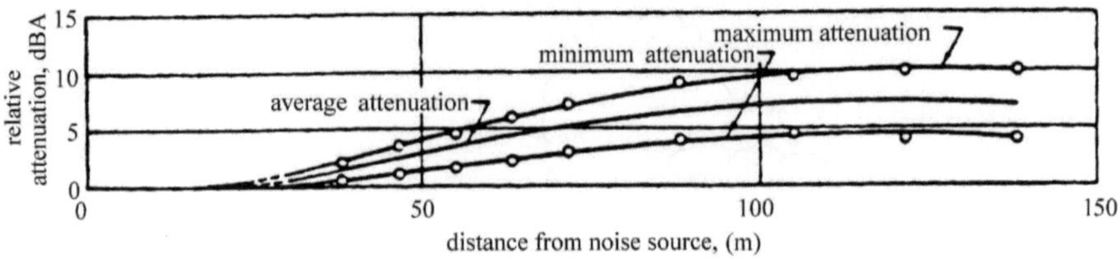

Fig. 2.10. Relationships between relative attenuation and distance from the noise source for different media in three characteristic situations presenting: *maximum attenuation* trees and corn, *average attenuation* gravel and *minimum attenuation* paving (Cook and van Haverbeke 1971)

2.2.2.2 Effect of Frequency

To identify and quantify acoustic features of interest for the studies related to noise control, a detailed analysis can be performed, using more or less sophisticated frequency analysis instruments which serve different purposes, such as: the assessment of the severity of an environment, the identification of system response properties and the identification of sources and transmission paths.

The acoustic signals can be steady-state, transient, stationary or nonstationary. Depending on signal complexity, the parameters calculated are different. For a steady-state stationary signal represented in amplitude-time coordinates, the mean value and the mean-square value, or the weighted average (which is a simple linear sum of values over a specific time interval), are calculated. The spectral functions provide a frequency decomposition of the signal values. The computation of rms values in one-third octave bands is widely used for frequency analysis of acoustical data. Much fine frequency analysis requires very complex calculation of frequency spectra, using fast Fourier transform algorithms.

The reader interested in more details related to data analysis is invited to study Piersol's (1992) chapter "Data analysis" in the book edited by Beranek and Vèr (1992). See also Goodfriend (1977), von Gierke et al. (1998) and Gygi et al. (2004).

2.2.2.3 Effect of Visibility

The effect of visibility in the forest on the attenuation of noise was and still is a very intriguing question from the beginnings of "forest acoustics". To estimate the density of tree belts, Eyring (1946) and Embleton (1963) proposed the parameter "visibility", defined as the distance at which an object is obscured by the vegetation. Eyring (1946) stated that attenuation is correlated with visibility inside the tropical rain forest and that it increases with increasing frequency.

As regards visibility, we note the results reported by Pal et al. (2000), performed in different forests that were intentionally planted or preserved around coal mines in India in order to protect the neighborhood from pollutants and noise. The resulting data set was used to derive a linear relationship of excess noise attenuation (dB) as a dependent variable, with the independent variables of number density, average height, canopy branch cover, trunk diameter and both vertical and horizontal light penetration. It was observed that light penetration (which depends reciprocally on leaf density and branches) is the most decisive parameter, while the average density (number of trees per unit area) has only a negligible effect. It was supposed that "sound waves propagate through the gaps between the trees even with maximum plantation density". It was stated also that meteorological effects were supposed to be negligible for this experiment.

More recently, Fang and Ling (2003) studied tree belts in Taiwan (*Ficus microcarpa, Podocarpus macrophyllus, Palaquium formosanum, Camelia japonica*, etc.) and noted a negative logarithmic relationship between the visibility, belt width and the relative attenuation determined "as the difference between the measurements on open ground and data from the tree belt which includes the effects of distance and vegetation" (Fig. 2.11). On the graph, four regions can be observed, noted A, B, C and D: region A – reducing noise in that the relative attenuation exceeded 10 dB, region B – between 10 dB and 6 dB, region C – between 6 dB and 3 dB and region D – less than 3 dB.

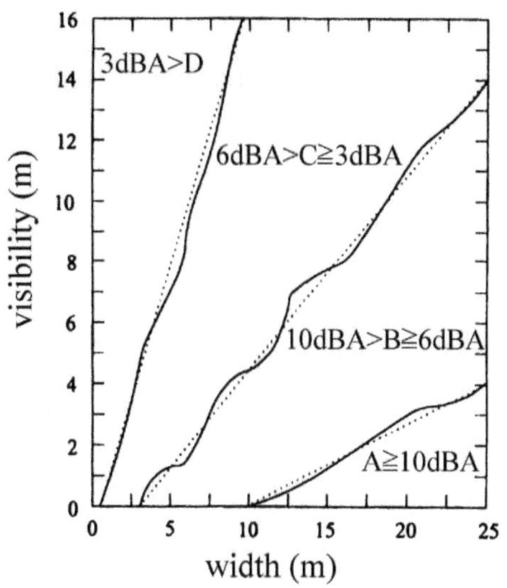

Fig. 2.11. Relative attenuation in a very large range (lower than 3 dB and higher than 10 dB) as a function of visibility and width of tree belt (Fang and Ling 2003). Reprinted with permission from Elsevier, copyright 2005

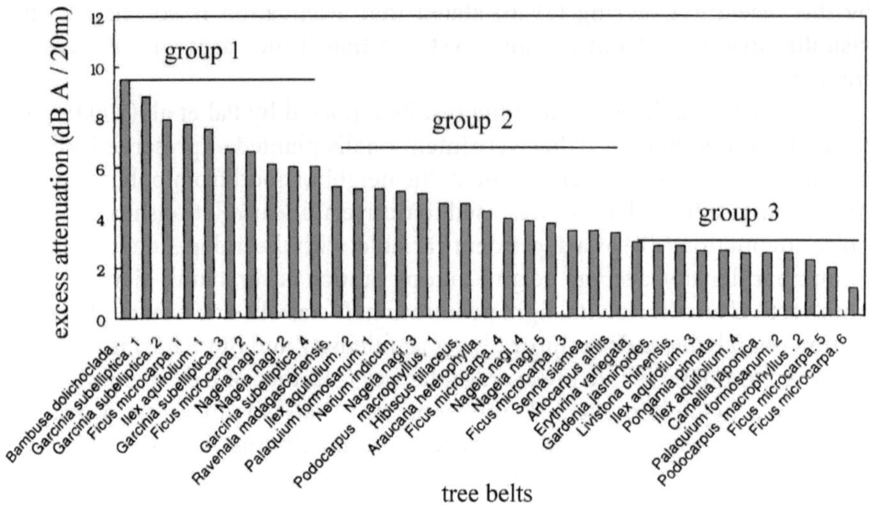

Fig. 2.12. Classification of tree belts following the excess attenuation value (Fang and Ling 2003). Reprinted with permission from Elsevier, copyright 2005

Figure 2.12 represents the excess attenuation of 35 tree belts classified in three groups:

- Group 1: "effective reduction" region for which the excess attenuation (dB/20 m) exceeded 6 dB, for a visibility less than 5 m.
- Group 2: "sub-reduction" region for which the excess attenuation is between 3.0 dB and 5.9 dB, for a visibility between 6 m and 19 m.
- Group 3: "invalid reduction" region, for which the excess attenuation is less than 2.9 dB, for a visibility exceeding 20 m.

From this study, it was recommended that a belt of trees and shrubs can reduce noise by 6 dB via suitable plantings and visibility (e.g. 1 m visibility and 5 m width, or 10 m visibility and 18 m width).

2.3 Summary

Noise in urban forest is produced by the sound field of different sources which can be detected in the surroundings. The acoustic intensity of this field is characterized by the following parameters: amplitude of disturbance, excess pressure, particle velocity, density change or corresponding change in refractive index, steady pressure on a surface due to the impact of sound waves, thermal changes produced by alternating compression and rarefaction, and the power which may be absorbed from the sound waves.

The specification of noise in physical terms depends upon its nature. One of the best representations is given by the spectrum. For noise measurement, three techniques are used: recording the wave-form to identify the disturbing frequency components, narrowband analysis and broadband analysis when determining the requirements for noise control. For most purposes, it is sufficiently accurate to use octave band analysis.

The measurable aspects of sound propagation in air can be described by the following parameters: sound pressure, speed of sound, sound power level, sound intensity level, sound pressure level, average sound level, averaged A-weighted sound level, day/night noise level and A-weighted noise exposure level. The system used for noise control contains three major components: the source, the path and the receiver, which are associated with emission, transmission and immission. The sound energy emitted by a noise source is transmitted to the receiver where it is immited.

The techniques and instruments for the measurement of the surrounding noise are determinants for noise control and abatement at any point in the acoustic field as a function of time and frequency. The environmental acoustical signal is processed with the sound-level meter. Outdoor measurements are mainly influenced by the distance effect, the frequency effect and the visibility, which is important mainly in tropical and subtropical areas.

3 Tree Characteristics and Acoustic Sensors

The foundation of forestry is closely related to forest mensuration, which is a keystone for obtaining quantifiable information when decision making for stands and individual trees. This chapter is devoted to tree characterization, using acoustic sensors for the following three topics: the morphological characteristics of trees, their mechanical and genetic characteristics related to timber quality, and tree characterization for sylvicultural practice. The studies of these aspects are required for the management inventory of forests, for timber volume estimation, for forestry planning and protection and for optimal timber management practices, such as pruning, thinning and logging operations.

3.1 Morphological Characteristics

Two main morphological characteristics are required for timber volume estimation and growth inventory in forest: the diameter and the height of a tree. It is generally accepted that the estimation and modeling of the height/diameter ratio for forest inventory require an important data base, which is time-consuming and is often performed using a small proportion of the sampled trees.

The diameter of the tree is currently measured with mechanical and electronic calipers, diameter sticks and tapes (McCornnell et al. 1984; Reynolds and Wilson 1989). This operation needs direct contact with the stem and is tedious over long periods. The development of ultrasonic and laser sensors has provided a noncontact method which allows the measurement of trunk diameters, the estimation of cross-sectional area and the collection and recording of data without manual entry.

The device developed by Upchurch et al. (1992, 1993), an ultrasonic caliper, uses high frequency waves of 25 kHz and two transducers to detect the presence of an object and to measure the time of flight (t) from the transmitter to the object and back to the receiver. The distance (d) is estimated with the formula $2t = d/v$, where v is the speed of sound in air, which is about 330 m/s and which needs to be adjusted for the temperature at which the measurement is taken. The distance between the tree and the sensor decreases as the diameter of the tree increases, when the transducers are aligned with the center of the trunk, using a stick. For diameters between 5.2 cm and 13.8 cm, the system has an accuracy of 0.05 cm in a temperature range between 0 °C and 34 °C. The major and minor axis of the trunk cross-section can be identified.

The development of mobile robotics and sonar modules combine ultrasonic (49 kHz) and laser techniques for measurements in dense forests and for all types of terrain and surroundings (Haglof 2003). The parameters measured are the distance and the angle to an optional part of the tree or to a reference point on the tree, or to the top and bottom of the tree, which allow calculation of the height and diameter of the tree at a reference point, as can be seen from Fig 3.1.

Taking into account the mechanical stability of trees against wind and storm, two other morphological parameters must be considered: root systems and crown characteristics.

Detailed information about root zone architecture and functionality can be obtained with a ground-penetrating radar technique which provides 3D images (Martinis 2002; Stokes et al. 2002; Nadezhdina and Cermak 2003) or by "air-spade" excavation, using a supersonic air stream which removes the soil around the dense network of roots (Rizzo and Gross 2000; Nadezhdina and Cermak 2003). The image obtained with a geo-radar operating at 450 MHz for about 30 m^2 ground surface is given in Fig. 3.2, in which only roots with a diameter greater than 20 mm are observed. The supersonic air stream is produced by a device related to a compressor delivering 0.8 m^3/s of air at a pressure of 0.6 MPa, "giving a stream with a speed of Mach 2".

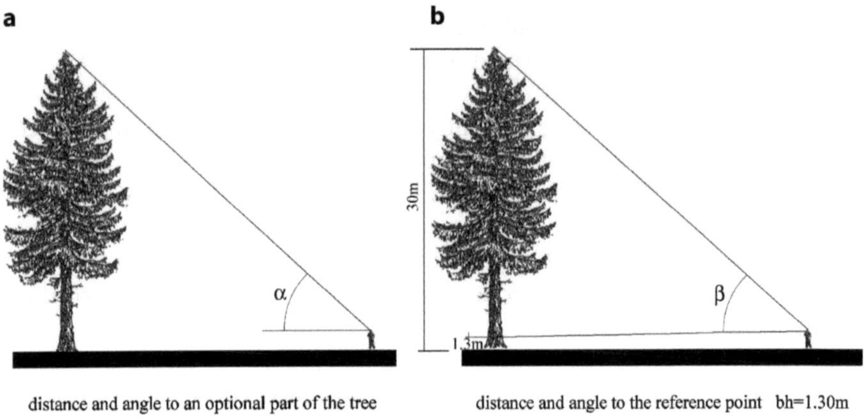

Fig. 3.1. Ultrasonic measurements of the height and diameter of trees. **a** Ultrasonic measurement of tree diameter when the transmitter and receiver are aligned with the center of the cross-section, using a hook device and engaging the device against the far side of the tree (Upchurch et al. 1992). Reprinted by permission of the American Society of Agricultural Engineers, copyright 2005. **b** Ultrasonic and laser measurements of the distance and angle to an optional point or to a reference point

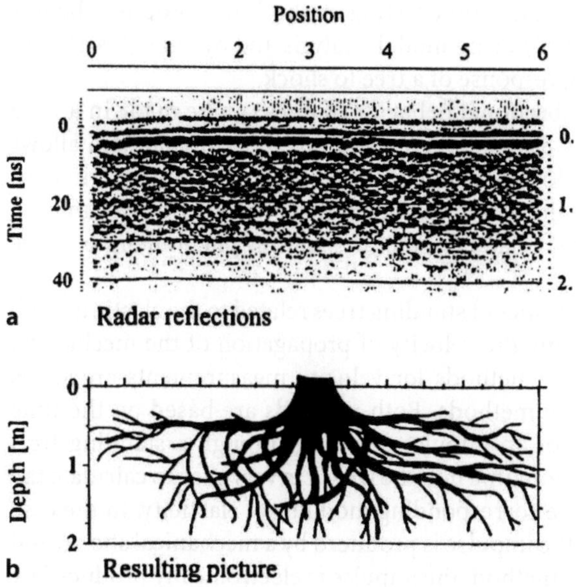

Fig. 3.2. Geo-radar image of the root system of a large oak tree (Nadezhdina and Cermak 2003). Reprinted by permission of Oxford University Press, copyright 2005. **a** Radar wave-path across the root system, **b** image of root system deduced from the radar data

Remote sensing for management inventory of forest condition and for forest protection needs information about canopy characteristics which can by studied using microwaves and radar techniques (McKerrow and Harper 1999; Bucur 2003).

3.2 Mechanical Characteristics

The stability against wind, storm, and snow or ice breakage of standing trees is directly related to the physical and mechanical properties of the fresh wood of the stem. For this purpose, static and dynamic acoustic methods were developed, concentrated mainly on determining the modulus of elasticity in the fiber direction. For static measurements, different devices were developed (Koizumi 1987, 1990; Launay et al. 2000; Takata and Teraoka 2002), based on the tree-bending test. The deflection caused by bending moment induced by the application of an external charge has been used for the calculation of stem Young's modulus E in the fiber direction (Lanbourg 1989). Brüchert et al. (2000) studied flexural stem variation by introducing defined mechanical parameters, such as: *structural Young's modulus*, which is the Young's modulus of the stem, *flexural stiffness*, which is the product between the Young's modulus and the axial second inertia momentum of the elliptical area of the stem ($I_a = 0.25\pi ba^3$, where a and b are the half diameters of the elliptical cross-area of the trunk) and the *global buckling* coefficient under the stem's own weight, which allows calculation of the tree safety factor under buckling.

Theoretical development of dynamic testing methods was proposed by Axmon and Hansson (1999, 2000) using modal analysis and by Ouïs (2001) with the analysis of the damping response of a tree to shock.

The description of the vibrational behavior of trees can be made in a time domain and in a frequency domain. Signal analysis in a time domain allows the determination of the velocity of propagation of the vibration and, in contrast, frequency analysis of the vibration allows the determination of damping characteristics of the signal related to the attenuation phenomena of wave propagation.

The mechanical characteristics of standing trees related to the elastic moduli can be achieved by measuring the velocity of propagation of the mechanical vibration. The most popular methods for velocity measurements are stress wave and ultrasonic velocity methods. Both methods are based on the time propagation measurement of an impulse traveling through a standing tree. Knowing the wave propagation time into the tree, it is very easy to calculate the propagation velocity and the corresponding modulus of elasticity. In the case of the stress wave method, the impulse is produced by a mechanical shock; and in the case of the ultrasonic method, the impulse is electronically produced by a piezoelectric transducer.

The measurement of the damping characteristics of the signal traveling through standing trees is much more complex and is used mainly for the detection of internal defects related to the presence of decay (Ouïs 2001) or the proportion of juvenile wood when the analysis of the dispersion of multiple elastic waveguide modes is performed (Laverty 2001). The method developed by Ouïs (2001) is based on the assumption that each tree has proper vibration characteristics, related to the reverberation time that can be defined together with a characteristic response spectrum (transfer function). This method was inspired by the oldest test performed by foresters when they “listen to the sounding trees”.

3.2.1 Devices and Instrumentation

Figure 3.3 shows the measurement device for the stress wave method. The stress wave is generated by a shock induced by a hammer. The receivers are two accelerometers, the first one located at point 1 for velocity measurements in a longitudinal direction and the second one located at point 2 for measurements of radial velocity. The stress wave frequency is in the range 12 kHz.

The distribution of the sensors for modal analysis is given in Fig. 3.4. The stress wave propagates from the emission point to points 1, 2, etc., through the cross-section, in a straight line. The surface wave propagates at the periphery of the trunk and is received at the same points 1, 2, 3, etc. The shock is produced with a standard hammer and is received by accelerometers. For stress waves, the time of flight increases continuously from the emission point to a maximum corresponding to point 6 and decreases symmetrically from position 6 to position 12. The propagation time of surface waves increases linearly from position 1 to position 12, covering the entire circumference. The velocity of the surface wave is constant at the periphery of the stem.

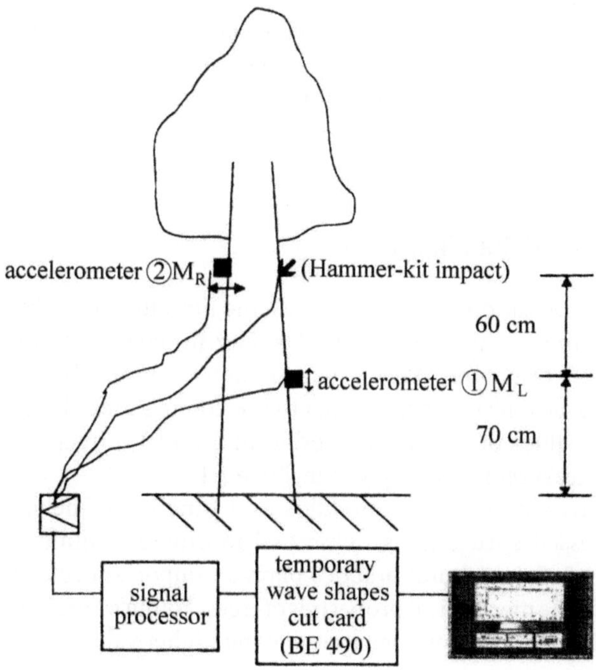

Fig. 3.3. Device for stress wave method on a standing tree (Chuang and Wang 2001). Reprinted by permission of the Japan Wood Research Society, copyright 2005

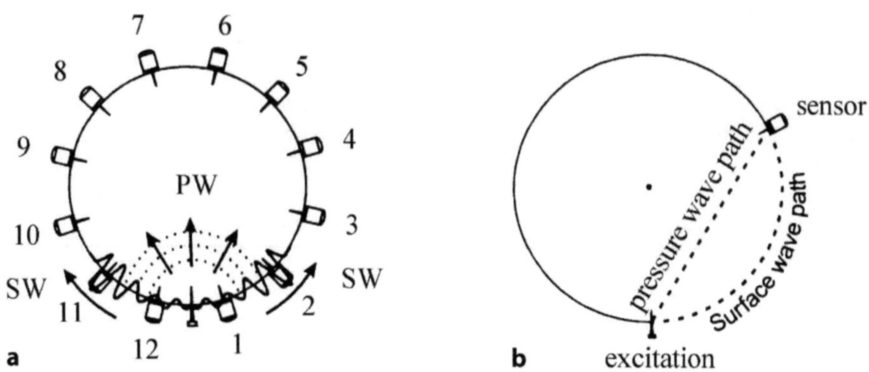

Fig. 3.4. Location of accelerometers for modal analysis (Axmon et al. 2004). Reprinted by permission of Oxford University Press, copyright 2005. **a** For stress wave (SW) velocity measurements, the shock is induced at point P. **b** Propagation path of surface waves

From modal analysis, temporal and spatial frequencies are estimated for each spatial mode shape of vibration, which is associated with the temporal frequency of the stress wave. The pattern generated by the surface waves is separable from the pattern generated by the stress wave. The discrepancy between the calculated and measured frequency then indicates whether the tested tree has internal defects or not.

The ultrasonic device utilizes two piezoelectric transducers (one for the emission and other for the reception of the signal), which are located in similar way as for the stress wave method. The frequency range is much higher than that used for stress wave measurements and can range between 20 kHz and 1 MHz.

3.2.2 Mechanical Characteristics of Standing Trees

For the mechanical characterization of standing trees, the literature available today mentions two parameters: the moduli of elasticity in axial (longitudinal, L) and in radial (R) directions. Very often the abbreviation MOE is advanced, which in mechanical terms corresponds to the Young's modulus in a longitudinal direction. Values of the elastic moduli in L and R directions derived from ultrasonic measurements are given in Table 3.1.

The variation in the structural Young's modulus as a function of height for dominant and suppressed spruce trees, submitted to different thinning regimes, is given in Fig. 3.5. Sylvicultural practice plays an important role in the differences observed in the mechanical properties of trees. Young's modulus decreases with stem diameter, the suppressed trees having higher values.

Table 3.1. Elastic moduli and ultrasonic velocities in L and R directions in Douglas fir trees (Bucur 1995)

Parameter	Unit	Pruned tree	Control tree
Velocity in R direction	(m/s)	1,589	1,272
Velocity in L direction	(m/s)	6,006	5,528
Density	(kg/m^3)	547	550
Modulus of elasticity in R direction	10^8 N/m^2	138.11	88.98
Modulus of elasticity in L direction	10^8 N/m^2	197.31	168.07

3.2.3 Detection of Internal Defects in Standing Trees

For the detection of internal defects in standing trees, two main groups of methods were developed: the ultrasonic velocity method (Bucur 1985; Leininger

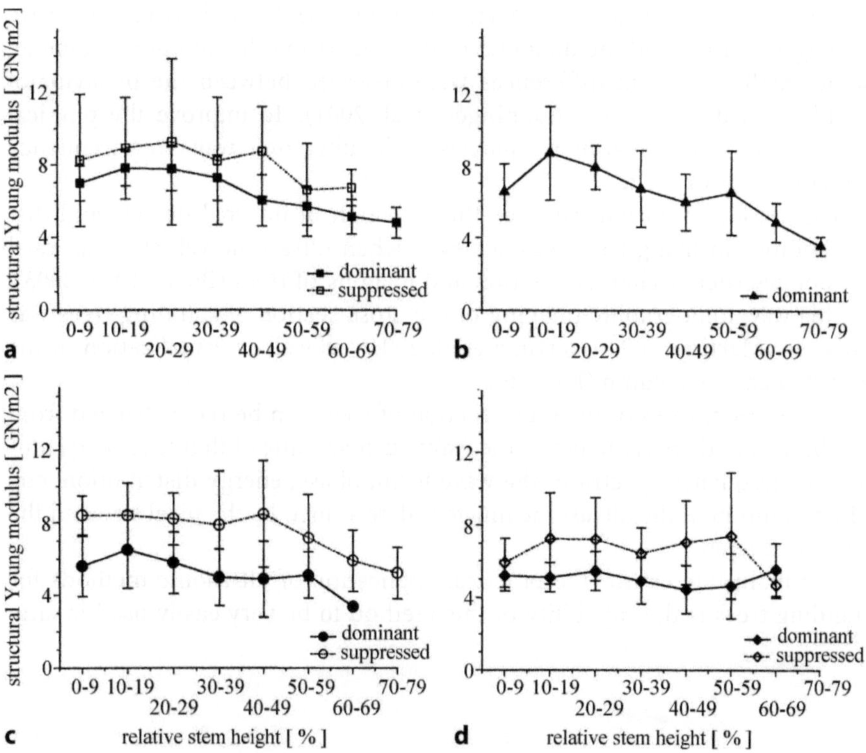

Fig. 3.5. Relationship between the structural Young's modulus and the relative stem height for dominant and suppressed spruce trees in four sites in Germany (Brüchert et al. 2000). Reprinted with permission from Elsevier, copyright 2005. *Filled symbols* represent dominant trees and *open symbols* represent suppressed trees. Sites: **a** wide spacing, **b** dense spacing, **c** initial wide spacing, **d** dense spacing, high amplitude

et al. 2001; Martinis et al. 2004) and the stress wave method (Ross 1985; Mattheck and Bethge 1993; Mattheck 1996), for which the impact response was analyzed either by modal analysis (Axmon et al. 2004), or by analysis of the damping response of the tree (Ouïs 2001) for internal decay detection in living trees or by exploration of the dispersion of multiple elastic waveguides in small diameter logs (Rizzo and Gross 2000; Laverty 2001) for the detection of the proportion of juvenile wood.

3.2.3.1 Ultrasonic Velocity Method

The development of an ultrasonic method for the detection of internal defects in trees used mainly two parameters: the time of flight of the ultrasonic signal and the ultrasonic velocity. First, only the time of flight of the ultrasonic signal was measured. Experiments reported a strong correlation between the wave propagation time and the diameter of the tree. Using this unique parameter, some small significant differences were observed between the behavior of healthy and decayed trees (Leininger et al. 2001). To improve the physical approach of signal treatment, analysis of the ultrasonic wave in a frequency domain was proposed.

The second important step for the detection of internal defects and discontinuities in living trees was achieved when ultrasonic velocity was used for nondestructive characterization and imaging of trees (Bucur 1995, 2003). In this way, tomographic imaging of the cross-section of standing trees was possible (Martinis 2002; Martinis et al. 2004), as well as visualization of the extent of any rot column (Fig. 3.6).

Ultrasonic images of the cross-section of trees can be reconstructed from all characteristic parameters of the wave such as: time of flight, velocity, amplitude, frequency spectra of the wave form, phase, energy distribution, etc. The resolution of the ultrasonic image is determined by the pixel size and the beam diameter.

The main interest for the practical application of ultrasonic methods for standing trees is the capability of this method to be very easily used in situ.

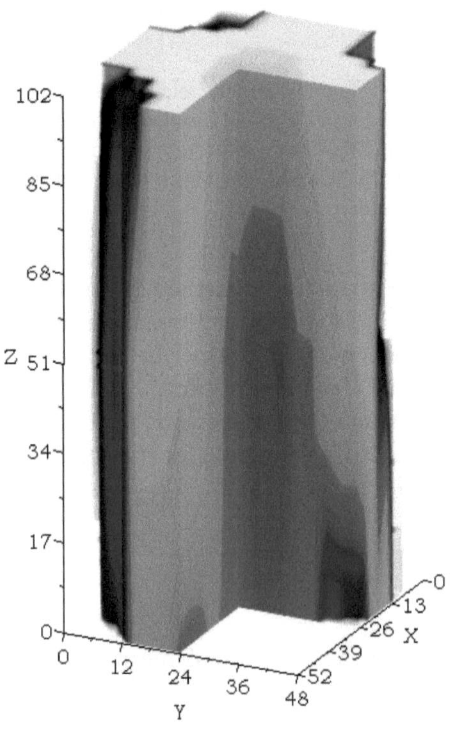

Fig. 3.6. Visualization of the extent of a rot column with ultrasonic tomography (Martinis et al. 2004). With permission from EDP Sciences, copyright 2005

The main disadvantage is the need for a good acoustic contact between the transducers and the bark or wood of the stem.

3.2.3.2 Stress Wave Method

During the past decades, the stress wave method was developed in the United States for the nondestructive estimation of the quality of wood products (Pellerin 1965; Ross 1985; Ross and Pellerin 1991, 1994; Wang et al. 2000; Pellerin and Ross 2002; Wang and Ross 2002). More recently, to evaluate the performance of the stress wave method through the detector rule for sound trees and decayed trees, Axmon et al. (2004) suggested the introduction of the parameter labeled *residual threshold,* which represents the difference between the measured frequency and the predicted, theoretical frequency of the ovaling mode for a supposed sound tree. From Fig. 3.7 (residual threshold versus surface wave velocity), it is easy to note that the sound trees are grouped in the central zone of the graph at ±46.9 Hz, while the decayed trees are located outside this zone. Successful classification within the major classes – sound and decayed trees – versus the double-sided residual threshold is given in Fig. 3.8. The success in identifying the sound trees is 74% for the residual threshold of 53 Hz.

Ouïs (2001) referred also to the ovaling mode of vibration of the tree stem, for which the wavelength is $2\lambda = 2\pi R$. He defined the *reverberation time* of the decay of the mechanical shock propagating through the tree, using an original

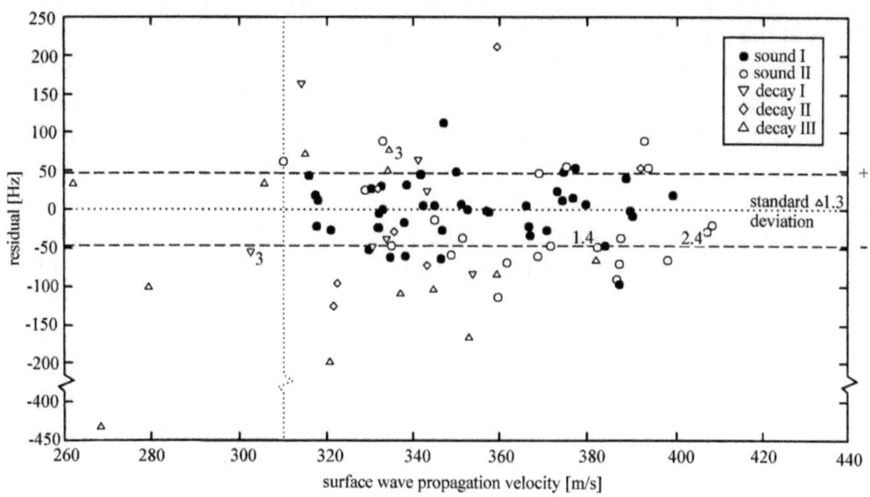

Fig. 3.7. Residual threshold (difference between measured and theoretical frequency) versus surface wave velocity (Axmon et al. 2004). Reprinted by permission of Oxford University Press, copyright 2005

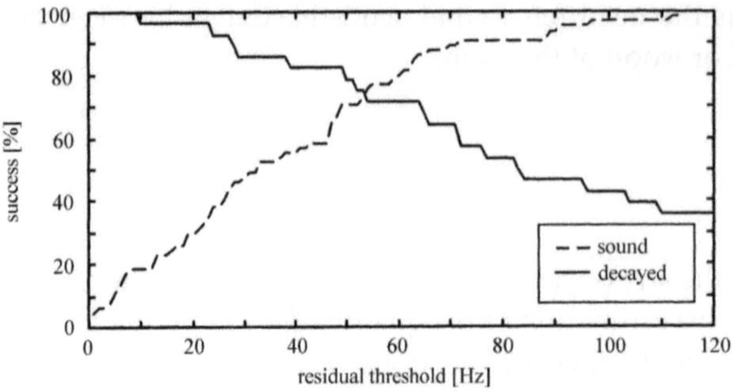

Fig. 3.8. Classification of sound and decayed trees as a function of residual threshold (Axmon et al. 2004). Reprinted by permission of Oxford University Press, copyright 2005

approach related to the analysis of the impulse response. First, the impulse response "is squared and then integrated backwards to yield the energy decay curve from which the reverberation time is calculated" as the time necessary for 10 dB decreasing of the level of the integrated impulse response. The impulse response in time and frequency domain of a sound and decayed (infested) tree is given in Fig. 3.9. An important discrepancy was observed between the velocity of impulse propagation in a sound tree (546 m/s) and a decayed tree (289 m/s). The spectrum analysis revealed harmonic components at different frequencies for sound and decayed trees, as for example for the sound tree at 1,710 Hz, 2,850 Hz, 4,170 Hz, etc. and for the decayed tree at 1,088 Hz, 1,920 Hz, 2,660 Hz, etc. This characteristic frequency distribution can be used for discriminating between sound and infested trees. The synthetic interpretation of all these experimental data is given in Fig. 3.9c. The energy decay curve versus time shows that the decayed (infested) tree lost much more energy during the decay of the impact vibration than the sound tree.

The increasing commercial interest for structural utilization of small diameter logs revealed the problems related to the estimation of the proportion and mechanical properties of juvenile wood. Laverty (2001) studied the ability to detect the proportion of juvenile wood in a log through exploration of the dispersion of multiple elastic waveguide modes propagating simultaneously. The developed model admitted that a log is composed of two concentrically cylindrical layers having a transverse isotropic symmetry and that the juvenile wood corresponds to the inner cylinder. The discrimination between the mature and juvenile wood can be made by studying the dispersion curves of vibrations propagating in the tree and by selecting the number of modes, the corresponding wavenumber of the signal and the shape of the waveguide modes. The combination of number of modes and shape of modes proved that the problem can be reduced to measuring the external diameter of the stem.

In the future, the development of a mobile device for practical application of this noninvasive technique in forests and saw-mills will be of the greatest interest for in situ appreciation of wood quality in a very big number of trees.

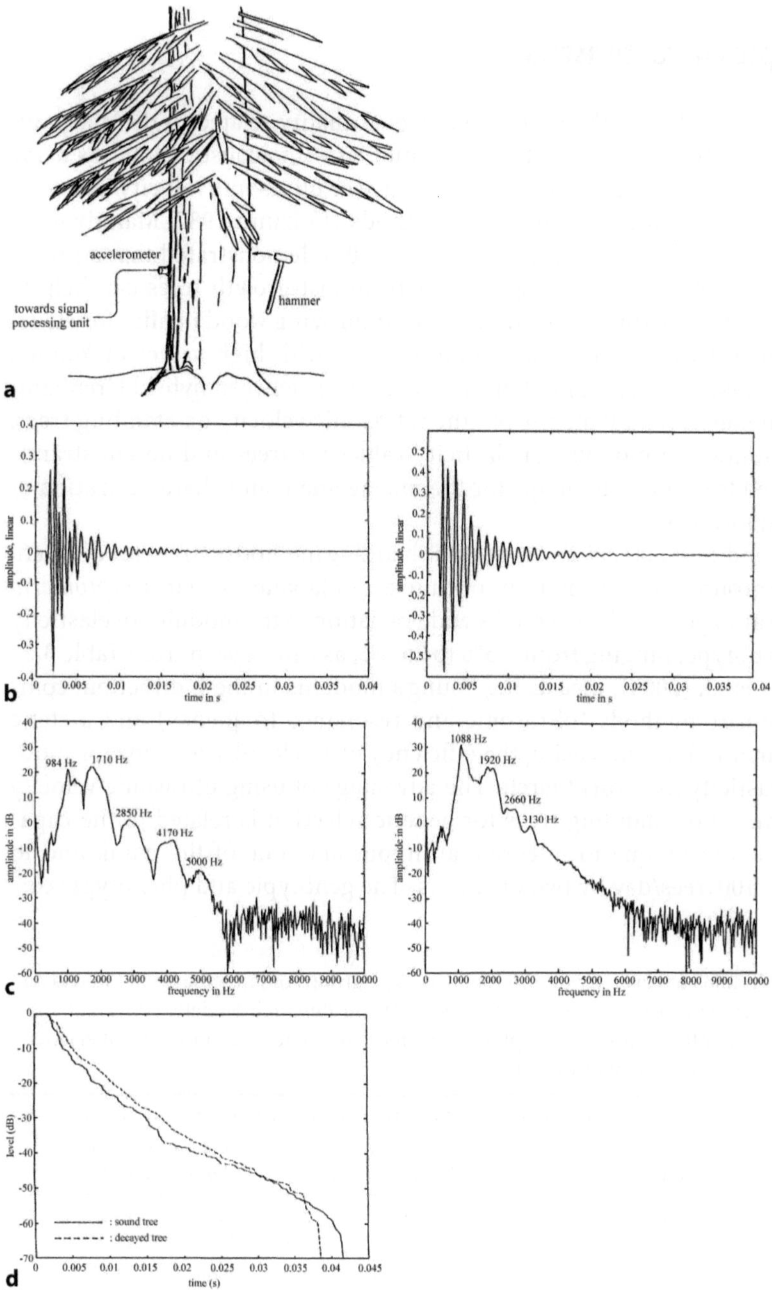

Fig. 3.9. The impulse response in time and frequency domain of sound and decayed trees (Ouïs 2001), with permission. **a** Time domain response, **b** frequency domain, **c** energy decay curve versus time, **d** energy decay curve of sound and decayed tree in time domain

3.3 Genotypic Characteristics

The genotypic effect of the variation of wood quality related to the mechanical characteristics of standing trees was studied with acoustic methods using resonance frequency FFT analysis, stress-wave or ultrasonic velocity measurements as well as with static bending methods (Koizumi 1987; Mamdy et al. 1999; Launay et al. 2000). Lindström et al. (2002) demonstrated that stiffness determination on small clear specimens from fast-growth trees can help to capture genetic breeding opportunities for improving wood quality for structural lumber, by selecting *Pinus radiata* clones with high values of Young's modulus. Nakamura (1996) noted that differences between hybrid larch families can be observed by measuring the ultrasonic velocity on standing trees. The distribution of modulus of elasticity values for trees in different strands can be used for producing maps for the management and characterization of individual forest sites.

Takata and Teraoka (2002), using static bending method on trees for different genotype groups, from plantations of cultivars of Japanese cedar (*Cryptomeria japonica*) at 19 years old, observed a wide variation in the modulus of elasticity of each genotype, ranging from 7.5% to 26.8%, as can be seen from Table 3.2.

Jacques et al. (2004), measuring Young's modulus in the L direction, compared acoustic methods (ultrasonic and resonance frequency) and a static bending method for increasing the efficiency of 16 clonal selections for modulus of elasticity for hybrid larch. The advantage of using ultrasonic velocity measurements on standing trees for genetic selection is related to the capability of this technique to integrate an important zone of the trunk and to test about 100 trees/day by two operators. The genotypic and phenotypic cor-

Table 3.2. Modulus of elasticity of different genotype groups of Japanese larch (Takata and Taraoka 2002); with permission from the Japan Wood Research Society, copyright 2005. Numbers in parentheses are the samples for the measurement of the modulus of elasticity of the trunk. CV Coefficient of variation

Genotype	Tested trees	Tree diameter (cm)		Tree height (m)		Trunk modulus of elasticity (GPa)	
	Number	Mean	CV (%)	Mean	CV (%)	Mean	CV (%)
A	31 (16)	17.4	23.1	10.8	15.8	5.53	12.2
B	26 (20)	18.5	14.4	11.4	13.4	5.38	11.2
C	15 (11)	17.5	20.7	11.3	14.3	5.68	11.0
D	9 (7)	20.2	16.3	11.5	17.5	4.81	7.5
E	8 (6)	17.6	27.4	10.8	22.8	5.69	26.8
F	6 (4)	17.8	13.8	10.7	7.3	5.39	18.8
Overall	108 (71)	17.9	20.1	11.0	15.3	5.42	14.5

relation coefficients between the modulus of elasticity determined with the ultrasonic velocity method, resonance frequency method and static bending range between 0.985 and 0.998.

The ultrasonic velocity measurements on standing trees used for the ranking of the 16 clones gives a lower estimate of heritability and genotypic than on small clear specimens. However, this small deficiency is compensated by the enormous cost reduction related to sample preparation and by the possibility of the increasing of selection intensity by performing direct measurements on standing trees.

In the future, genetics has an important role to play in forestry for fast-grown plantations. Acoustic methods (Huang et al. 2003) can be used to grade standing trees or sawn logs according to their suitability for structural lumber or the pulp industry. The performance of these methods depends on the wide range of fundamental wood characteristics (cellulose microfibril angle in wood cells, fiber length, etc.) interlinked with acoustical properties. "In long time perspective, there are substantial benefits in using acoustics in tree breeding to screen for candidate trees with superior wood properties" (Huang et al. 2003).

3.4 Sylvicultural Practices

The influence of different spacing on standing tree quality of Japanese cedar (*Cryptomeria Japonica*) was studied by Chuang and Wang (2001) with stress wave and ultrasonic methods. The experiments evaluated the wood quality through the measurement of the modulus of elasticity of 47 old standing trees growing in five plantation sites (denoted A, B, C, D, E) with different spacing,

Table 3.3. The effect of spacing on some acoustic and mechanical characteristics of trees grown in plantations at different spacings. Data from Chuang and Wang (2001), with permission from the Japan Wood Research Society, copyright 2005

	Spacing	Diameter	Density	Velocity (m/s)				Modulus of elasticity (10^8 N/m^2)			
				Stress wave in axis		Ultrasonic wave		Stress wave		Ultrasonic wave	
	(m)	(cm)	(kg/m^3)	L	R	L	R	E_L	E_R	E_L	E_R
A	1 × 1	29.0	386	3,210	1,719	3,600	1,720	101	29.9	117	32.9
B	2 × 2	29.5	397	3,520	1,830	3,810	1,800	129	35.7	147	33.0
C	3 × 3	34.1	409	3,200	1,777	3,440	1,881	94.3	32.2	114	30.5
D	4 × 4	37.4	431	3,230	1,777	3,430	1,770	98.7	26.8	117	29.3
E	5 × 5	40.0	442	2,900	1,810	3,280	1,740	82.7	33.2	89.3	28.2

ranging from 1 × 1 m to 5 × 5 m. In each type, the trees were classified in three classes: S – superior growth trees, M – medium growth trees and P – poor growth trees. Velocities in L and R directions were measured with stress wave and ultrasonic methods. Table 3.3 summarizes the experimental data. The diameter range is between 20 cm and 40 cm.

Figure 3.10 shows a histogram of the tree diameters at breast height as a function of the site. Statistical analysis of data with multiple new-ranged Duncan's test showed that significant differences were observed between the

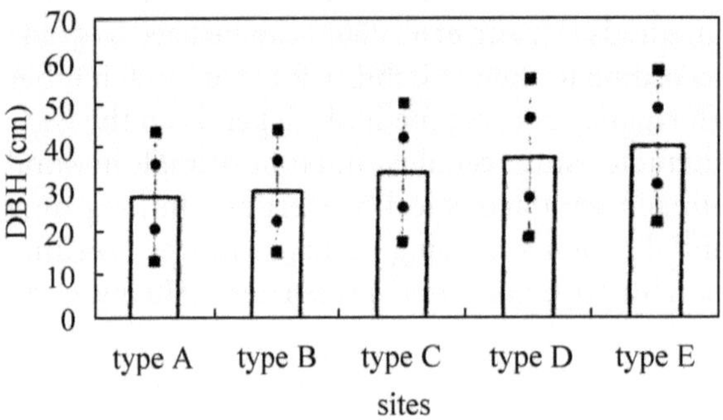

Fig. 3.10. Histogram of tree diameter at breast height for five plantation sites (Chuang and Wang 2001); with permission from the Japan wood Research Society, copyright 2005

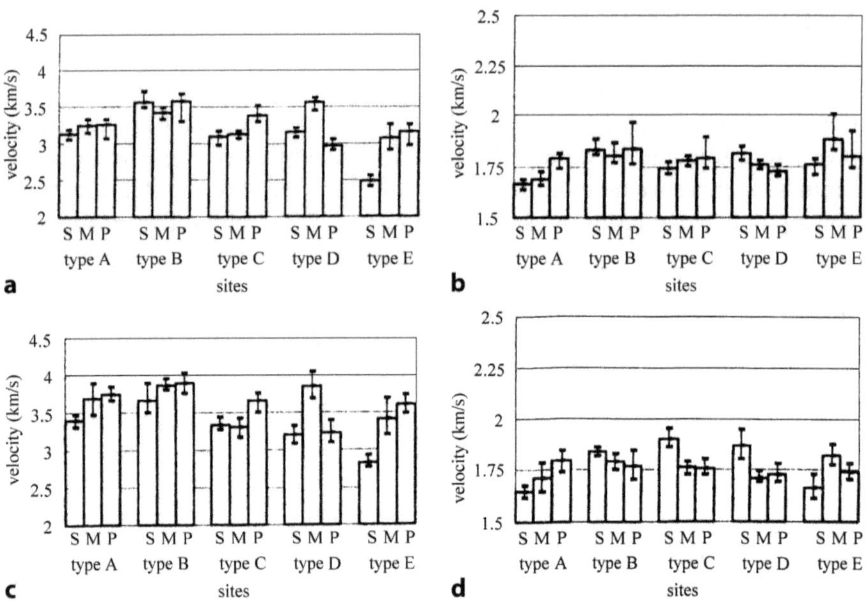

Fig. 3.11. Velocities measured with two techniques as a function of growth conditions (Chuang and Wang 2001); with permission from the Japan Wood Research Society, copyright 2005. **a** Longitudinal velocity of stress waves, **b** radial velocity of stress waves, **c** longitudinal velocity of ultrasonic waves, **d** radial velocity of ultrasonic waves

Table 3.4. Multiple new-ranged Duncan's test for the modulus of elasticity (mean and standard deviation, in 10^8 N/m^2) measured with stress wave and ultrasonic velocity methods. Data from Chuang and Wang (2001), with permission from the Japan Wood Research Society, copyright 2005

Parameter	Type				
Plantation site	Type E	Type C	Type D	Type A	Type B
E_L stress wave	82.70 ± 6.14	94.30 ± 4.18	99.70 ± 10.11	101 ± 1.13	128 ± 6.06
Significant difference at $P < 0.05$	Ns	Ns	Ns	Ns	–
Plantation site	Type D	Type A	Type C	Type E	Type B
E_R stress wave	26.80 ± 2.43	29.90 ± 2.65	32.20 ± 9.00	33.20 ± 2.18	35.70 ± 1.81
Significant difference at $P < 0.05$	Ns –	Ns –	Ns Ns	Ns Ns	– Ns
Plantation site	Type E	Type C	Type D	Type A	Type B
E_L ultrasonic waves	99.30 ± 11.70	114.00 ± 8.98	117.99 ± 3.87	117.00 ± 6.51	147.00 ± 10.07
Significant difference at $P < 0.05$	Ns –	Ns –	Ns –	Ns Ns	– Ns
Plantation site	Type E	Type C	Type D	Type A	Type B
E_R ultrasonic waves	28.20 ± 2.23	29.30 ± 2.43	30.50 ± 1.03	32.90 ± 1.70	33.00 ± 1.32
Significant difference at $P < 0.05$	Ns	Ns	Ns	Ns	Ns

diameter of trees belonging to sites C, D and E. But no differences were found between the diameters of trees in sites A and B.

From Table 3.3 and from Fig. 3.11, it can be noted that the values of ultrasonic velocities are slightly higher than those determined with the stress wave method. The differences are probably due to the frequency differences between the two methods. Meanwhile, a decrease can be observed in velocity value and modulus of elasticity with increased spacing between trees. This is probably due to the effects induced by increasing annual ring width, the ratio of earlywood to latewood density, etc. These parameters are directly related to the plantation spacing. As can be seen from Table 3.4, a multiple new-ranged Duncan's test for the modulus of elasticity values with ultrasonic and stress wave methods, shows that the plantations can be classified as follows: type B > type A > type D > type C > type E. The plantation type B is characterized by the highest values of the modulus of elasticity and it can be admitted that rapidly growing trees with significant spacing (5 × 5 m) have low strength properties.

Based on these observations, it was stated that a classification of stands can be established and optimal management practices (pruning, thinning, etc.) selected.

3.5 Summary

Tree characteristic (diameter, height, etc.) measurements using acoustic sensors are required for the management inventory of forests, for timber volume estimation and for forestry planning and protection. The main morphological characteristics for grown inventory are the diameter, the height and the height/diameter ratio. The ultrasonic caliper allows noncontact measurement of trunk diameter and estimation of the cross-sectional area and height of the tree. The development of mobile robotics and sonar modules combines ultrasonic and laser techniques for measurements in dense forests and for all types of terrain and surroundings. Beside the trunk, both the root system and crown characteristics must be taken into account for the mechanical stability of trees against wind and storm. The geo-radar technique provides 3D images, while the "air-spade" technique uses a supersonic air stream to prevent any damage during excavation of the roots. The mechanical characteristics expressed by the moduli of elasticity can be measured with the ultrasonic velocity and stress methods. For both methods, the time of propagation of a shock is measured. In the case of the ultrasonic method, an electronic pulse is used, while in the case of the stress wave method, a mechanical shock is produced by a hammer. The measurement of the damping characteristics of the signal traveling through the standing tree is much more complex and is used mainly for the detection of internal defects. Appropriate devices and instrumentation are described. Mechanical characteristics are also used for the study of genotypic characteristics of different clones. Sylvicultural practices, illustrated by the influence of spacing on the wood quality of standing trees is demonstrated through the measurement of the modulus of elasticity.

4 Noise Attenuation with Plant Material

4.1 Physical Aspects of Noise Attenuation by Vegetation

Outdoor sound propagation may range from a relatively simple to a very complex phenomenon, depending upon the nature of the source and the distribution of the surrounding area. To understand outdoor sound propagation and the transmission of sound and noise with plant material, it is necessary to study the complex acoustic climate of a plant community. The influence of a particular soil surface on sound propagation and the influence of overgrowing plant organs like stems, branches, twigs and foliage on the sound field inside vegetation are equally important. It is generally admitted that plants can attenuate sound by reflecting and absorbing energy in the viscous and thermal boundary layers near the plant surface, or by internal damping of sound-driven oscillations of branches or stems (Embelton 1963; Kragh 1979; Aylor 1977; Martens 1980; Bullen and Fricke 1982).

Figure 4.1 synthesizes the factors influencing noise attenuation in a forest stand through absorption, dispersion, reflection and refraction. Noise attenuation in its totality is composed of normal attenuation and excess attenuation.

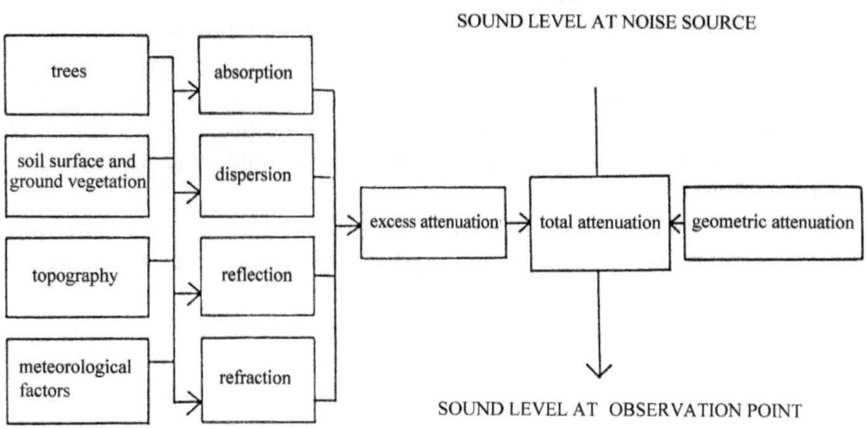

Fig. 4.1. Contribution of several factors (trees, soil, topography, meteorology) to total noise attenuation in a stand (Kellomäki et al. 1976)

Normal attenuation is due to spherical divergence and air absorption (Herrington 1974; Embelton 1996). Normal noise attenuation increases with distance, producing the well known "distance effect". As noted by Embelton (1966): "at twice the distance from the source, the surface of the wave front is four times as large, and the sound pressure level decreases by 6 dB. For a line of sources (e.g. a line of cars along a road), the sound wave spreads cylindrically in two dimensions. In this case, the sound pressure level decreases by 3 dB per doubling of distance, which is the half-rate for spherical spreading".

Furthermore, reflection, refraction, scattering and absorption effects due to any obstruction (barriers, ground, vegetation, trees, hills, etc.) between noise source and receiver result in excess attenuation (Fang and Ling 2003).

The attenuation of outdoor propagation sound (Bies and Hansen 1996) can be determined by the following four steps:

- the determination of the sound power level (L_W) of all sources;
- the calculation of the individual components of excess attenuation for all sources;
- the computation of the resulting sound pressure level at selected points in the environment for each of the individual sources;
- the computation of the predicted sound pressure level produced by all of the individual sources at selected points in the environment.

Attenuation by a tree belt, defined as the intensity at its far edge, relative to the intensity I_0 which is the intensity of *a plane wave incident* on it from one side, depends firstly on the scattering cross-section and absorbing cross-section of individual trees, secondly on the number of trees per square meter and thirdly on insertion loss. If the first and second factors can be easily understood, the insertion loss needs more comment, related to the nature of the ground – soft or hard. If the ground is acoustically soft, due to interference between direct and ground-reflected waves, significant attenuation of low frequencies may occur in the absence of vegetation. Over hard ground, the phenomenon is different; and the sound level can be locally increased due to consecutive interference. This effect would be destroyed by vegetation, since the phases of waves arriving at a point on the far side of the belt would be random (Bullen and Fricke 1982).

If the *incident wave is not plane* but arises from a source at a finite distance from the belt, it would normally undergo spherical spreading, which would also be disturbed by the presence of vegetation.

Scattering phenomena in the horizontal plane are very different from scattering in the vertical plane. In the horizontal plane, the wave undergoes a number of scattering events before leaving the vegetation; and it is possible to imagine that a wave which is scattered at a large angle to the horizontal will leave the vegetation without further scattering. In the vertical plane, scattering through the "top" of the vegetation is different from that on the "bottom" near the ground.

Empirical relationships were established between excess attenuation, frequency and distance of sound traveling through a heavily wooden area. In 1961, Hoover, cited by Bies and Hansen (1996), recommended the following equation:

$$A = 0.01 r f^{\frac{1}{3}} \tag{4.1}$$

where r is the distance (m) and f is the frequency (Hz).

As regards the attenuation of a tree belt and vegetation of different sizes – length and width – it appears that about 3 dB excess attenuation (over the infinitely long belt case) may be gained by making the belt approximately as wide as it is deep (Bullen and Fricke 1982). Scattering from the side of the belt can of course be observed, but much greater gains in attenuation than this will probably not be possible, since diffraction around the belt will become more important when the belt becomes narrower.

4.2 Ground Attenuation

A typical forest surface is a multi-layer structure, containing much decayed plant matter, such as leaves, needles, branches, decayed trunks and loose soil. Forest soils may be classified, in most cases, as acoustically soft (Martens et al. 1985). Three type of acoustic wave can propagate through or near the ground: plane waves, spherical waves and surface waves. If the propagation phenomena of plane waves and spherical waves are relatively easy to understand, the surface wave propagation requires some explication. The surface wave has a direction of propagation, which is parallel to the porous ground surface, and a direction of polarization associated with the elliptical motion of air particles as the result of combining motion parallel to the surface with that normal to the surface in and out of the ground.

When sound travels from the source to the receiver, close to the ground, an interaction is observed between the direct sound and the sound reflected from

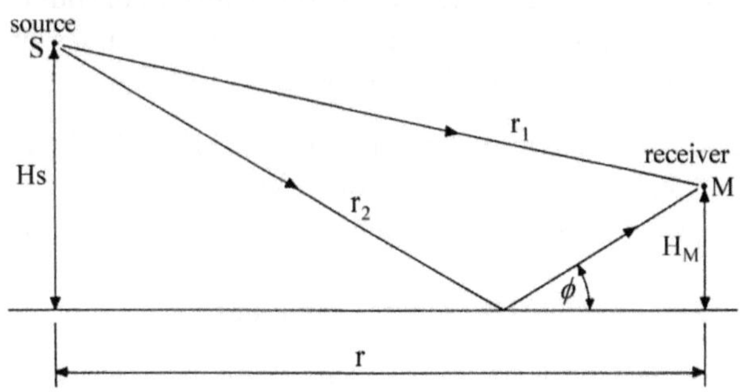

Fig. 4.2. Sound propagation path above a flat absorbing ground (Attenbourg 1988). Reprinted with permission from Elsevier, copyright 2005

the ground, as can be seen in Fig. 4.2. As noted by Reethof et al. (1977), the effect of the ground on sound absorption is due to the important porosity of the forest surface and to the interference between the ground-reflected wave and the direct wave. This interference forms a coherent source some distance above the ground, producing a "cancellation" effect. The ground-reflected wave has to travel a longer distance than the direct wave, so that there will be a location at a certain distance from the source where the two waves are exactly at opposite phases, bucking one another. Because of the porosity of the ground, the reflected wave, as it leaves the ground, will be at a somewhat different phase than the incident wave at the ground surface. Compared with the incident wave, the phase of the reflected wave is retarded by the delay, due to the increased distance travelled by the reflected energy and of course by the soil porosity. If the interaction between the sound and the ground is to be measured, the first request is to note the dependence of experimental data on source – receiver geometry.

The relationships between the shape and length of ground on sound propagation and the attenuation spectrum were reported over the years (Ingård 1953; Martens 1977; Reethof et al. 1977; Martens et al. 1985; Wempen 1986; Attenborough 1988; Embelton 1996).

The most important ground physical parameters studied were:

a) *the porosity* (expressed in %).
b) the *normalized characteristic impedance* (defined as the ratio of the pressure and normal velocity at the surface of a semi-infinite medium divided by the characteristic impedance; product of density and velocity). The real part of ground impedance is called "resistance" and the imaginary part is called "reactance" (Attenborough 1992).
   Wempen and Mellert (1990) suggested an empirical model for ground impedance (Z), written in terms of frequency (f) and *relative admittance* (β):

$$\beta = \frac{1}{Z} = 0.012 + 0.006\mathrm{i} + 60f \exp\left[-\mathrm{i}\arctan\left(\frac{0.003}{f}\right)\right] \tag{4.2}$$

c) the *flow resistivity*, σ, when knowledge of the propagation constant of sound within the ground layers is required. In the next lines, several values of effective flow resitivity, for different grounds, are given:

   - the pine forest $\sigma_e/1{,}000 = 7.5$ in a frequency range of 0.05 kHz to 7.5 kHz;
   - new snow $\sigma_e/1{,}000 = 5.5$ in a frequency range of 0.1 kHz to 5.0 kHz;
   - wet sandy loam $\sigma_e/1{,}000 = 4{,}546$ in a frequency range of 0.1 kHz to 2.0 kHz;
   - grassland $\sigma_e/1{,}000 = 3{,}000$ at 2,500 Hz (Attenbourg 1988);

   More detailed data on flow resistivity and porosity are given in Table 4.1.

Table 4.1. Measured values of flow resitivity (R_s) and corresponding porosity for different soils (Martens et al. 1985). Reprinted with permission from the Acoustical Society of America, copyright 2005

Surface	Soil layer	Flow resistivity (R_s; 10^3 Pa s m^{-2})	Porosity (%)
Lawn	R, layer with roots	237 ± 77	50.5 ± 9.3
Bare sandy plain	A, mineral soil	366 ± 108	36.2
Soil layer with roots	R, layer with roots	114 ± 53	55.2 ± 4.5
Grass covered soil	R, layer with roots	189 ± 91	–
Beech forest	L, litter layer	22 ± 13	82.3 ± 1.9
Pine forest	L, litter layer	9 ± 5	67.5 ± 4.1
Mixed forest with beech and pine	L/F, litter/fermentation layer	13.3 ± 3.0	–
	F/H, fermentation/humus layer	52 ± 24	76.3 ± 3.2
	H, humus layer	210 ± 93	–
	A, mineral soil layer	102 ± 60	54.7 ± 6.8
Mixed deciduous forest with oak and beech	L/F, litter/fermentation layer	30 ± 31	–
	H, humus layer	375 ± 69	84.6 ± 5.6
	A, mineral soil layer	540 ± 92	51.5 ± 4.8

d) *the ground surface admittance*:
The modelling of sound propagation over a finite impedance ground proposed by Attenborough (1988) takes into consideration the pressure reflection coefficient which, for a plane wave is:

$$R_p = |R_p| \exp(i\varphi) \tag{4.3}$$

where φ represents the phase change on reflection.
The total pressure at the receiver is given by:

$$P_t = P_d + R_p P_r \tag{4.4}$$

where P_d is the direct contribution and P_r is the speculary reflected contribution.

As noted by Attenborough (1988) "for a given source – receiver geometry, the two contributions P_d and P_r will lead to minima in the total pressure at frequencies where they interfere destructively or, in other words, when the phase difference between them is an odd number of φ radians (180°). The phase difference is caused both by path length difference and by the phase change φ on reflection at the ground, so r_1 is the length of the direct ray from the source to receiver, and r_2 is length of the reflected ray."

The condition for minimum in the total pressure is:

$$(2n+1)\pi = \frac{2\pi}{\lambda}(r_2 - r_1) + \phi \tag{4.5}$$

where λ is the wavelength, or the frequency (f_m) is:

$$f_m = c_0[(2n+1)\pi - \phi]/[2(r_2 - r_1)] \tag{4.6}$$

The extreme ground conditions are, for an acoustically *hard boundary*, for which we have $\phi = 0$ and a pressure-release boundary, for which $\phi = \pi$. For the former case, the frequency of the first (f_h) and subsequent minima is given by the equation:

$$f_h = c_0(2n+1)/[2(r_2 - r_1)] \tag{4.7}$$

where $n = 0, 1, 2, 3, \dots$ and c_0 = velocity of sound in air.

The frequencies of the associated pressure minimum (f_{pr}) for a pressure release boundary are given by the equation:

$$f_{\text{pr}} = nc_0/(r_{\acute{e}} - r_1) \tag{4.8}$$

If the *ground is porous*, it will have finite impedance and φ will be non-zero $\phi \neq 0$.

From (4.7), it can be stated that the resulting pressure minimum will occur at lower frequencies than those predicted for an acoustically hard boundary. The first minimum, called by Attenborough (1988) the *ground effect dip*, depends strongly upon the acoustical characteristics of the ground and relatively little upon the source – receiver geometry. In this theoretical approach, it was stated that the sound wave is plane and the sound field was produced by a point source.

In practical outdoor situations, the waves are spherical. In this case, two main aspects must be considered. Firstly, in the near field, the pressure due to a point source above an absorbing plane is inversely proportional to distance from the source and, secondly in the far field, the pressure is inversely proportional to the square of the distance.

As regards the relationships between sound frequency and distances, it can be noted that, for "frequencies less than 300 Hz and for ranges greater than 50 m over typical grassland, a surface wave, decays principally as the inverse square root of the horizontal range and exponentially with height above the surface. At grazing incidence, the condition for its existence is simply that *the imaginary part* of the ground impedance (*the reactance*) is greater than the *real part* (*the resistance*)" (Attenborough 1992).

For grass-covered and forest soils (Martens et al. 1985), the real part of the acoustic impedance is relatively independent of frequency, while the imaginary part strongly decreases with frequency, as can be seen from Fig. 4.3, in which different type of forests (mixed deciduous, pine forest, beech forest) and soils (intact soil, sandy soil, ivy underground) are studied.

A deep insight on the properties of forest soils has been obtained by performing laboratory tests on soil samples (Reethof et al. 1977), using an adapted

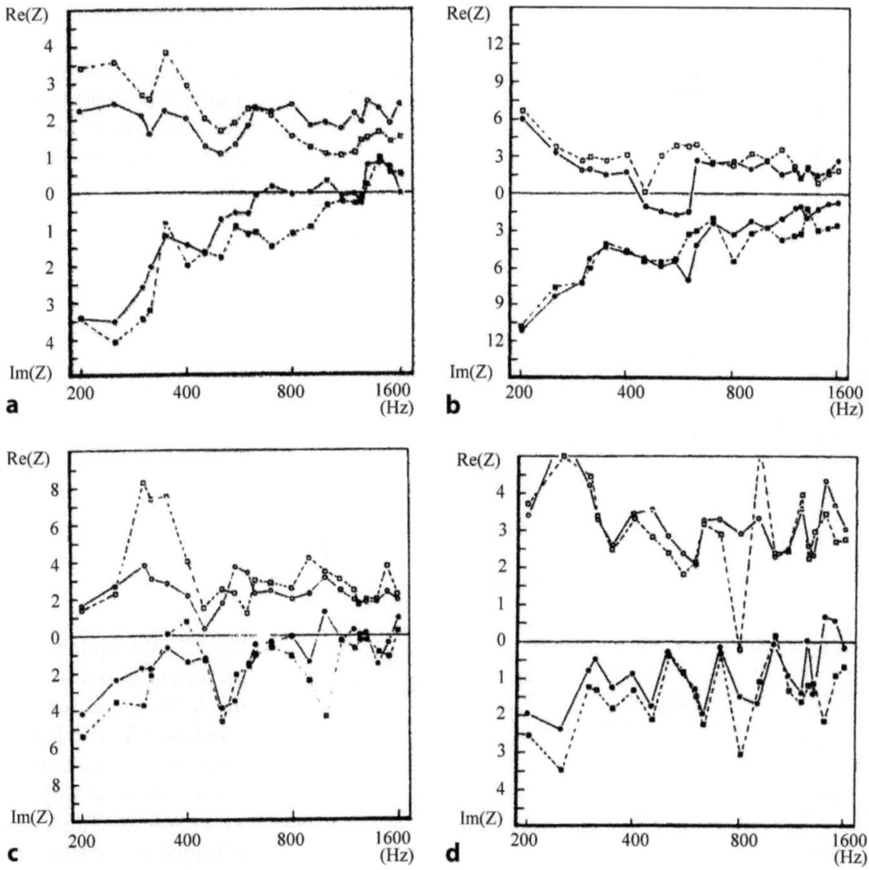

Fig. 4.3. Real and imaginary parts of the acoustic impedance versus frequency in different forests (Martens et al. 1985). Reprinted with permission from the Acoustical Society of America, copyright 2005. **a** Pine forest compared with intact soil (*circles*), **b** fir forest, **c** beech forest, **d** elm forest

standing-wave tube. Figure 4.4 shows a typical curve of the absorption coefficient (α) of a sample of forest soil as a function of frequency. A primary peak was observed at 250 Hz and a maximum value at 1,000 Hz, between them a dip low value at 500 Hz. Note the very good repeatability of the acoustical measurements.

Ground porosity variation induced by the contribution of the leaf layer in a deciduous forest is shown in Fig. 4.5. The absorption coefficient increases with increasing frequency. Compared with grass, the leaf layer determines about 20% of the increase in the absorption coefficient.

Analyzing previous data and comparing them with the noise spectra of trucks and automobiles, which are fairly flat with peaks in the 125-Hz octave

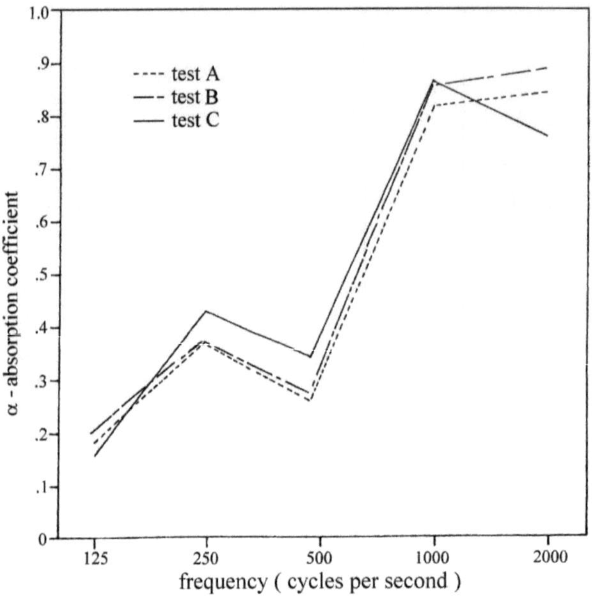

Fig. 4.4. Absorption coefficient of a normal incidence sound by a sample of forest floor, as a function of frequency (Reethof et al. 1977)

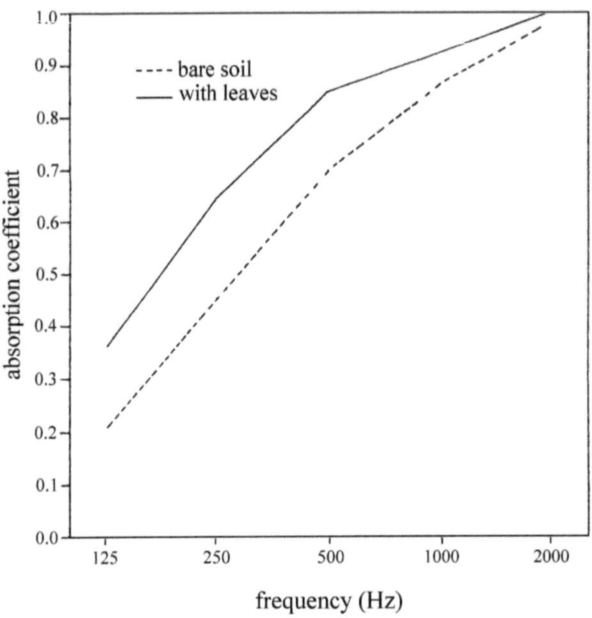

Fig. 4.5. Absorption coefficient at normal incidence sound for a soil sample with leaf litter from a deciduous forest compared with a bare soil, as a function of frequency (Reethof et al. 1977)

band, it can be noted that increasing absorption at frequencies higher than 500 Hz, allows one to imagine that several hundred meters of tree belt width are required to produce an important reduction in the A-weighted sound levels from the traffic.

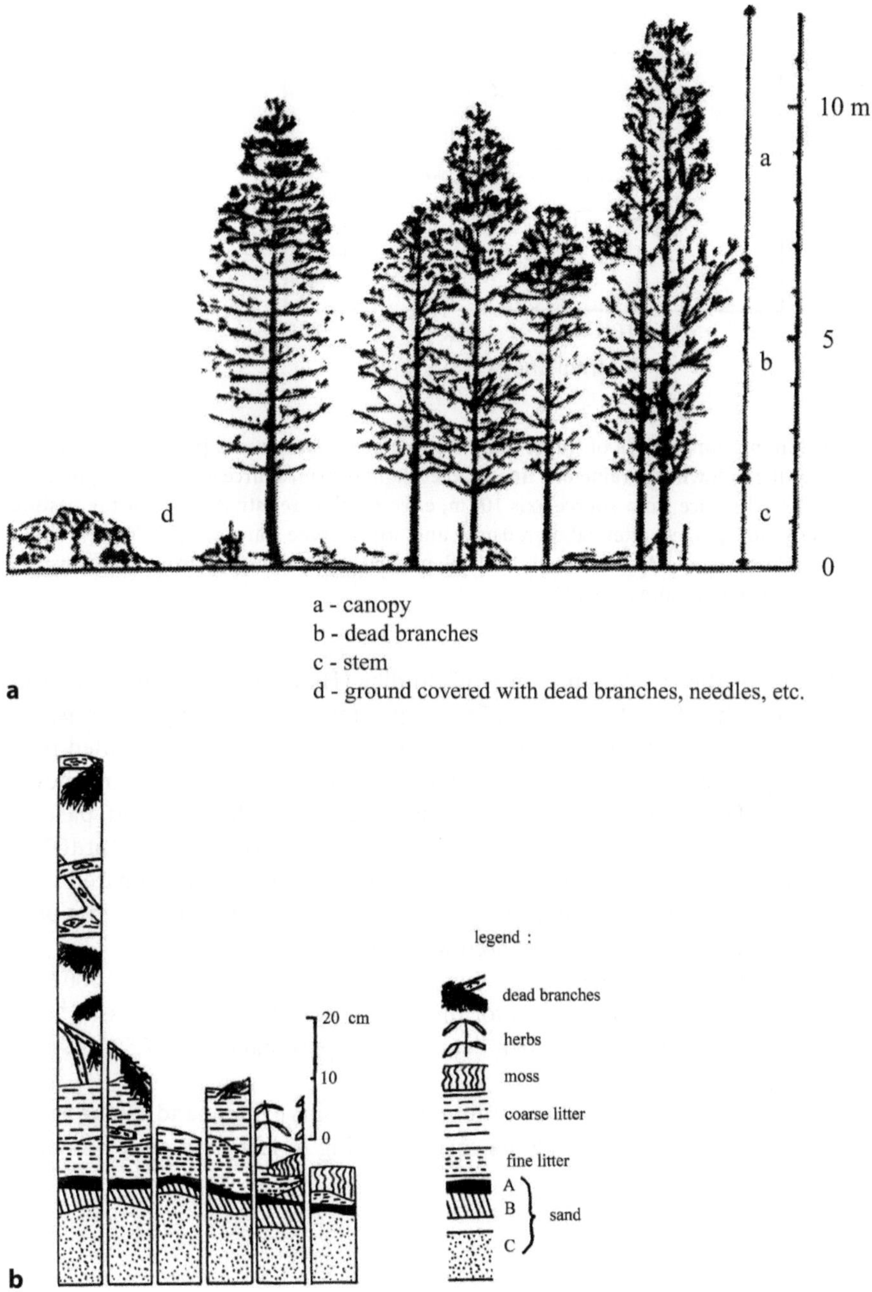

Fig. 4.6. Typical structure of a *Pinus nigra* monoculture stand locate on a soil including dead and living covering material (Huisman and Attenborough 1991). Reprinted with permission from the Acoustical Society of America, copyright 2005. **a** Profile of the vegetation structure, **b** soil structure

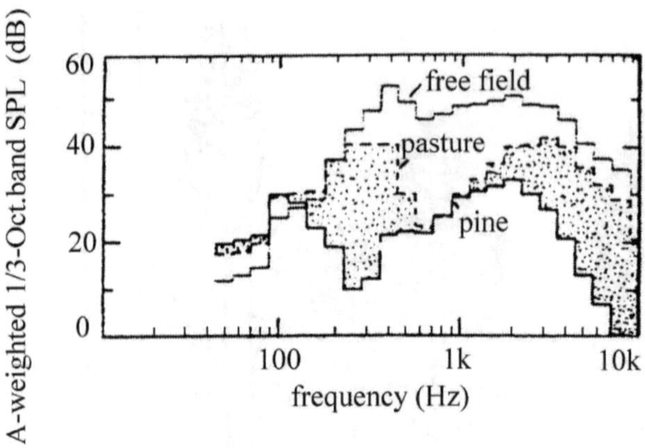

Fig. 4.7. Immission spectra of traffic noise for free field, pasture and pine stand for traffic noise for the following parameters: line source height 0.75 m, source length 600 m, receiver height 1 m, distance from source axis 100 m, effective flow resistivity (σ_e) for the pasture 125,000 N s m^{-4}, excess attenuation α 0 m^{-1}; and, for the pine stand σ_e 7,500 N s m^{-4} and α 25 m^{-1} (Huisman and Attenborough 1991). Reprinted with permission from the Acoustical Society of America, copyright 2005

Results of theoretical and "in situ" studies (Huisman 1990; Huisman and Attenborough 1991) of the effect of forest ground on the A-weighted imission level of road transmission noise, on a typical configuration of a planted pine forest (Fig. 4.6a), with a characteristic soil stratification (Fig. 4.6b) are shown in Fig. 4.7. This figure displays three immission spectra for free field, pasture and pine stand, calculated from a line source with road traffic situated at 100 m. The pine stand spectrum is lower for almost all frequencies, compared with pasture or free field. The total A-weighted immission spectrum in this pine stand is reduced by 9.9 dB, compared with pasture.

The soft forest floor has a big influence on low-frequency noise (< 500 Hz). The effective flow resistivity (σ_e) for the pasture was $\sigma_e = 125{,}000$ N s m^{-4} and the excess attenuation was $\alpha = 0$ m^{-1}; and for the pine stand, $\sigma_e = 7{,}500$ N s m^{-4} and $\alpha = 25$ m^{-1}.

In the studied pine stand, the ground effect seems evident and can be easily observed from the spectrum zone corresponding to low-frequency propagation.

4.3 Scattering by Trees

A simple calculation of the wavelength of a sound wave of 1,000 Hz frequency interfering with trees in a forest shows that the wavelength is comparable with tree diameter (e.g. $\lambda = 33$ cm for a sound velocity of 330 m/s). The incident acoustic waves are partially reflected and refracted, producing a typical scattering phenomenon, as shown in Fig. 4.8. The acoustic scattering and attenuation of sound are studied mainly along a line between a source and a receiver. The branches and the foliage partially scatter the incident acoustic energy to the side and backwards, producing a shadow zone behind the vegetation. The canopy of deciduous trees attenuates the incident noise. Plants in general and trees in particular can attenuate the sound by reflecting and absorbing energy in the viscous and thermal boundary layers near the plant surface or by internal damping of sound-driven oscillations of branches or stems (Aylor 1972a, b, 1977).

Scattering effectiveness is consistent with the geometry of scatterers such as trunk, branch and leaves (Huisman 1989). The bigger the scatterer, the lower the frequency at which the scattering phenomenon becomes effective. Scattering increases with frequency and the transmission path become more and more complex, producing absorption of acoustic energy. At low frequencies, this phenomenon is absent, because the wavelength is large compared to the diameter of trunks and branches; and the acoustic energy is transmitted easily. The propagation of sound through a large number of scatterers (trunks of

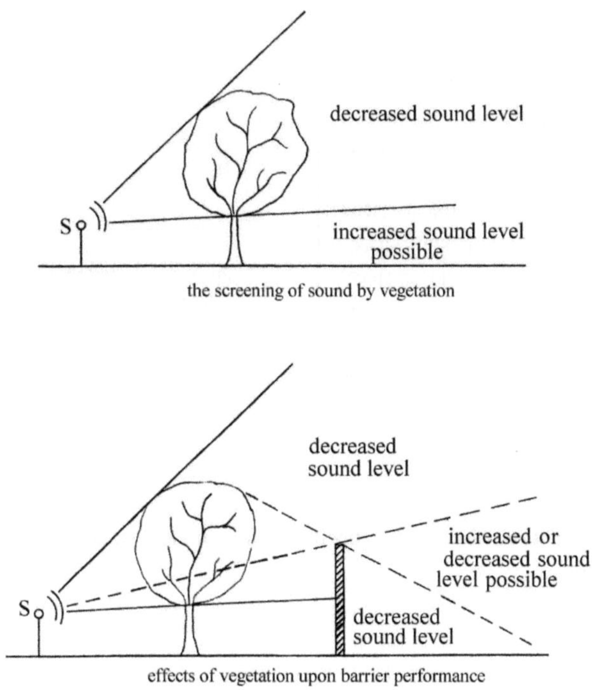

Fig. 4.8. Screening of sound by a tree (Lyon et al. 1977)

trees), in a first approximation, can be treated as a classic diffusion problem if the depth of the tree belt is large and the absorption is relatively low (Bullen and Fricke 1982). It was deduced that about 3 dB excess attenuation may be gained if the belt is as wide as it is deep. It was noted (Huisman 1989, 1990; Huisman and Attenborough 1991) that the interaction between trunk scattering and ground effect is much more complicated than the diffraction theory and more sophisticated modelling is necessary for a complete understanding of the experimental data produced by in situ measurements. Modelling the complex shapes of trees requires a reduction to simple shape components, such as cylinders, planes or spheres, for which analytical solutions for sound scattering are available, avoiding the application of numerical techniques that are expensive.

Figure 4.9 synthesizes the main dendrological and physical characteristics of the stand effecting excess attenuation in a tree belt. These characteristics are: the biomass of the stand, the structure of the stand in a horizontal plane (size and shape of the canopy) and the quality of the surfaces (size and shape of leaves and needles, soil). These characteristics allow admitting that mixed stands composed of coniferous and deciduous trees and bushes would be the most effective for noise attenuation.

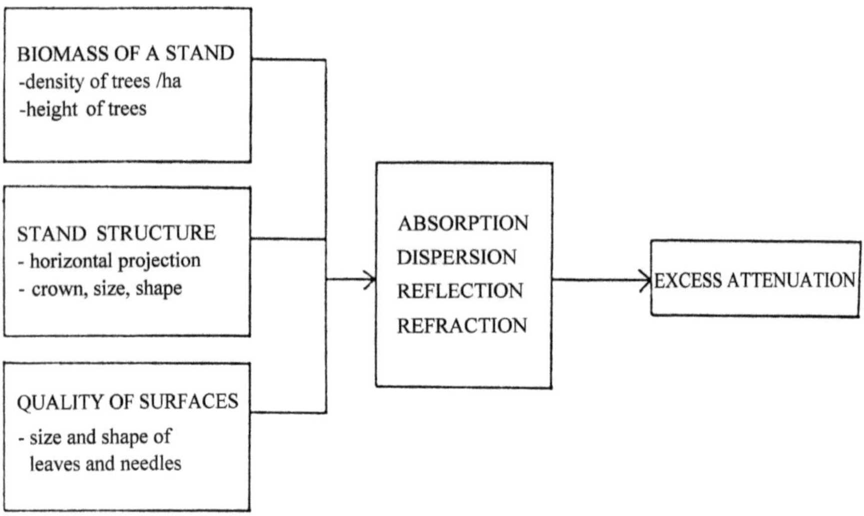

Fig. 4.9. Excess attenuation, absorption, dispersion, reflection, refraction and stand characteristics (Kellomäki et al. 1976)

4.3.1 Scattering by Stems

Because of the high complexity of acoustic scattering phenomena, Rogers et al. (1990) proposed theoretical studies, in an anechoic chamber, using models (wooden cylinders with limbs) to reproduce tree architecture (Fig. 4.10). It was noted that, at a fixed frequency, the optics scattering approximation can be applied to this acoustical study, if the illumination of the surface of the scattering solid by the source is correct and if this surface is the principal contributor to the total scattering.

Supposing a point source with acoustic strength A, placed at 1 m from the cylindrical specimen. The insonification of the cylinder $A_{(r)}$ can be calculated using the spreading factor:

$$A_{(r)} = A/r \exp(-ikr) \tag{4.9}$$

where r is the distance from the source to the point of interest on the cylinder and k is the wave number.

The back-scattered sound intensity at any point in space is found interacting over the illuminated zone of the cylinder, which is visible to the receiver. Each surface element becomes a source characterized by its amplitude, phase and geometric spreading factor.

Rogers and Lee (1989) analyzed the case of scattering from short cylinders (Fig. 4.11) normally illuminated and at 15° with respect to normal to the cylinder axis. As can be seen from this figure, the scattering pattern varies significantly with scatter angle; and the scattering is most important when the incident angle with respect to the cylinder axis equals the scattering angle. It was stated that these patterns are the radiation patterns of a line source with appropriate length and intensity. The maximum back-scattered acoustic signal occurs for signals incident normal to the axis of the cylinder. When this is not the case, the signal is considerably weaker.

Figure 4.12 displays the signals at the microphone and the back-scattered signal measured in an anechoic chamber for a trunk of 1 m length and 5 cm diameter, supporting six limbs of 0.4 m length and 2.5 cm diameter, as shown

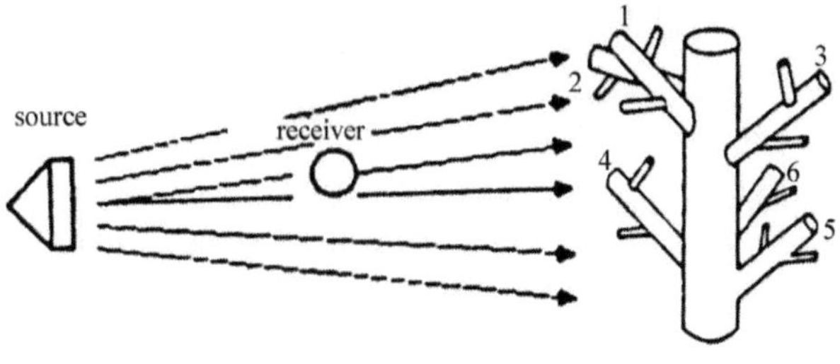

Fig. 4.10. Tree simulation for back-scattering measurements with cylindrical samples with limbs (Rogers et al. 1990)

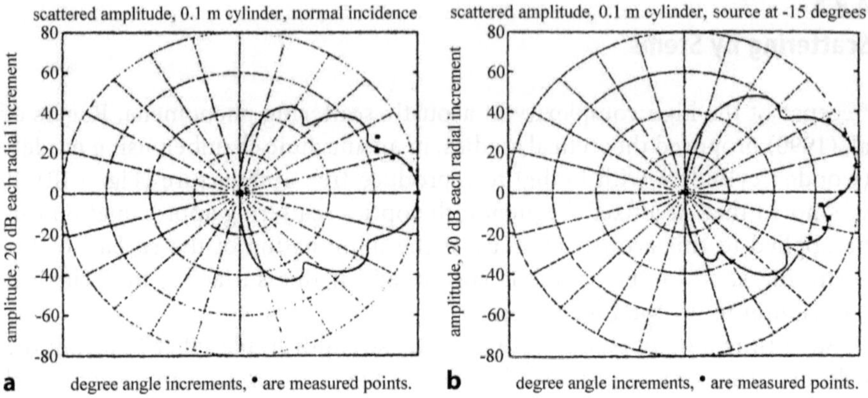

Fig. 4.11. Scattered amplitude from a cylinder of 0.1 m length. **a** source at normal incidence; **b** source at 15° (Rogers and Lee 1989)

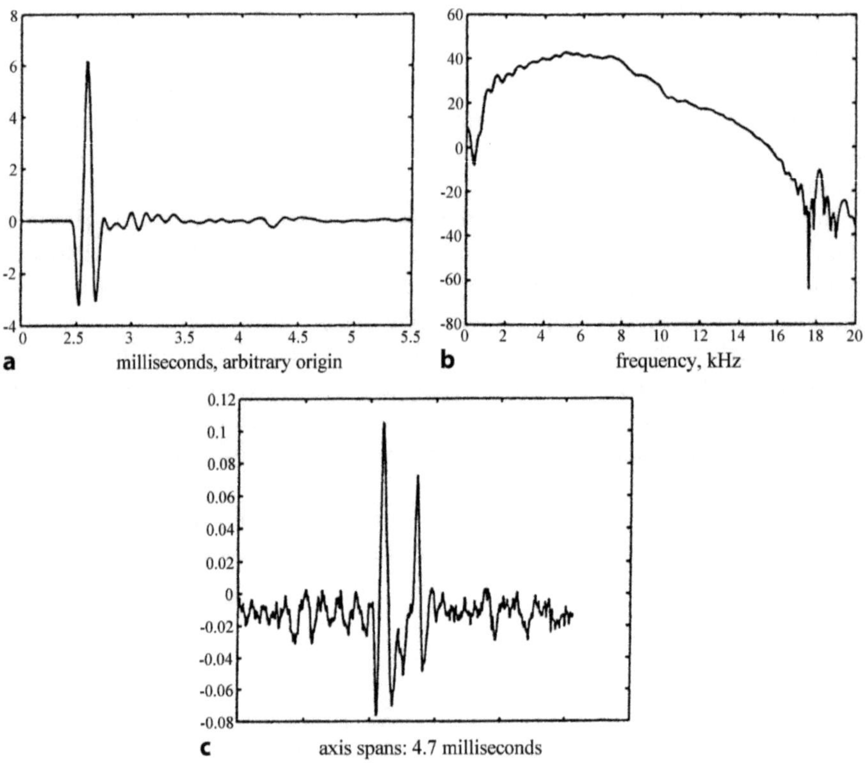

Fig. 4.12. Comparison between the emission signal and the back-scattering signal from one sample, in anechoic chamber (Rogers et al. 1990). **a** Direct signal at the microphone (amplitude in V); **b** corresponding spectrum (dB); **c** the back-scattering signal from the tree in a single direction (amplitude in V)

previously in Fig. 4.10. The useful bandwidth of the direct pulse was between 1 kHz and 12 kHz.

The next step in this approach was to study the back-scattering signals from tree samples arranged in a "grove" (Fig. 4.13). The corresponding spectra are given in Fig. 4.14. The frequency domain response of a single artificial tree from the grove is shown in Fig. 4.14a. and corresponds to the range 1–18 kHz. The responses of the nine artificial trees from the grove are shown in Fig. 4.14b, c. Both time and frequency domain signals are very complex, showing a high variability, which can be explained by the multiple returning pulses from different trees and from different zones of individual trees.

As far as the experimental data in the anechoic chamber with artificial trees seems to be coherent, it is therefore natural to consider the study of a natural tree belt. In this case, the ground adds more complexity to the interpretation of the experiments. Rogers et al. (1990) used a Freon horn source and a sound level meter located at 40 m and 60 m along a line normal to the edge of tree belt; and there was some cultural noise, automobile traffic and a temperature of 11 °C. Unfortunately in this report, there was no information about tree species and other practical parameters. Short blasts of the horn and the corresponding back-scattered signals were recorded (Fig. 4.15). The direct pulse spectrum is shown by the upper line, with the useful bandwidth between 0.5 kHz and 5.0 kHz. A big variability in the amplitude of the scattered signal is observed. The region between 2,500 Hz and 3,000 Hz displays less apparent attenuation than the region between 1,300 Hz and 1,800 Hz and between 3,300 Hz and 3,600 Hz. The spectral amplitude of the scattered signal is 30 dB below that of the direct signal from the tree belt. "If one assumes geometrical spreading, while ignoring ground effect, and uses the edge of the woods to approximate the spreading effect of the scattered signal, one would predict 14 dB reduction for the signal of a perfect back scatterer" (Rogers et al. 1990). The estimated scattering cross-section was 16 cm, which probably roughly corresponds to the tree diameter. In the case of the grove of trees, the ground effect does not exist and a simple geometric spreading predicts 15 dB attenuation.

Attenuation measurements in two pine plantations (labeled good and poor) were reported by Leonard and Herrington (1971). The dendrological charac-

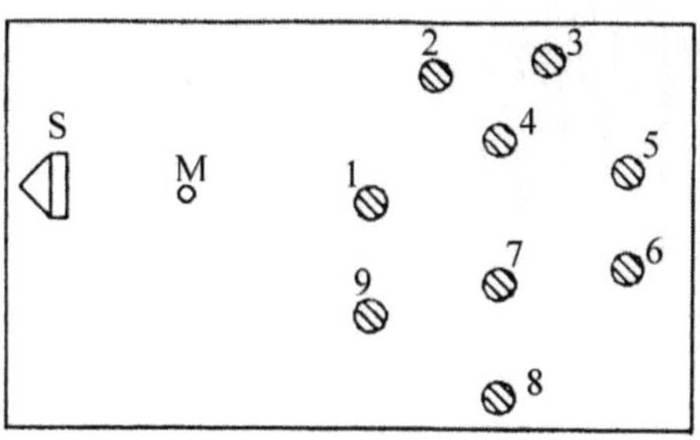

Fig. 4.13. Tree samples arrangement in a "grove" in an anechoic chamber for the simulation of a tree belt composed of nine trees. The source *S* and the microphone *M* is behind the trees (Rogers et al. 1990)

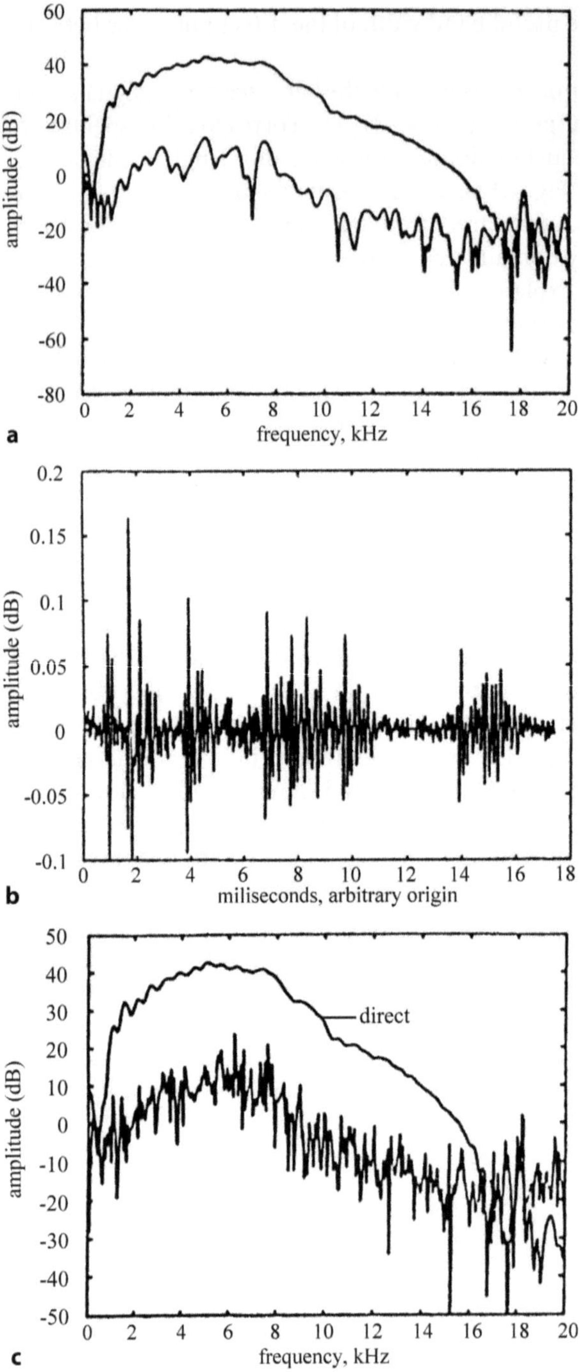

Fig. 4.14. "Grove" Spectra in frequency range 0–4,000 Hz (Rogers et al. 1990). In **a**, **c**: *upper* spectrum is direct signal, *lower* spectrum is scattered signals. **a** Spectrum of a single tree compared to direct spectrum (*upper*). **b** Synthetic scattering from nine trees grove in time domain. **c** Spectrum of the nine trees compared with direct signal spectrum

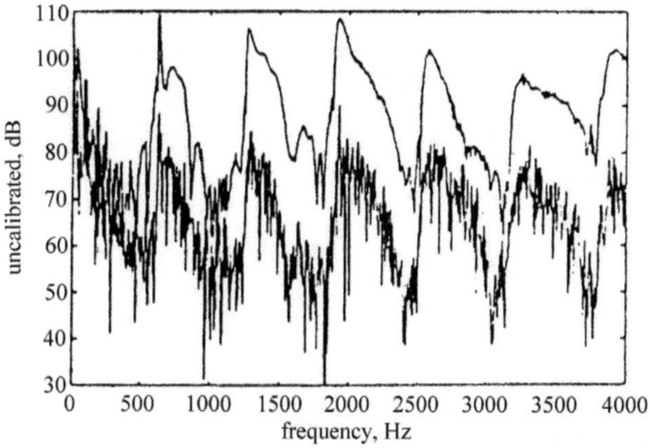

Fig. 4.15. Spectra for direct and scattered sound measurements on tree belt (Rogers et al. 1990). *Upper* spectrum is direct signal, lower spectrum is scattered signals

teristics of the trees are given in Table 4.2. The stem maximum diameter was between 11.2 cm and 12.7 cm and the maximum height between 12.6 m and 16.2 m. The sound pressure level was measured at different frequencies (125, 250, 500, 100, 200 Hz) and at different distances (2.6, 16.5, 33.0, 66.0, 82.0 m).

The excess attenuation displayed in Fig. 4.16 was calculated for different frequencies, with (4.10):

$$SPL_{r_2} - SPL_{r_1} = 20 \log \left(\frac{SPL_{r_1}}{SPL_{r_2}} \right) + A_{\text{excess}} \tag{4.10}$$

where r_1 and r_2 are the distances from the source.

The first term of this equation accounts for the reduction in the SPL in a free field, free from a boundary, for which a loss of 6 dB was measured. The second term includes the attenuating effect of atmospheric absorption, wind turbulence, temperature gradient, ground effect and trees. Very small differences were observed between the poor and good sites at 500 Hz. The "good site" produced more attenuation between 125 Hz and 500 Hz, while the "poor site" produced more attenuation between 500 Hz and 1,000 Hz.

Bullen and Fricke (1982) reported measurements in a reverberant room, on five young trees, about 2 m high. The absorption of the normalized cross-

Table 4.2. Dendrologic characteristics of the pine stand (Leonard and Herrington 1971)

Stand	Stem diameter (cm)		Height (m)		Height to live crown	Basal area
	Maximum	At 1.20 m	Maximum	minimum	(m)	(m^2/ha)
No. 1, good	12.7	10.9	16.2	13.4	8.4	35.0
No. 2, poor	11.2	7.4	12.6	7.9	4.9	16.6

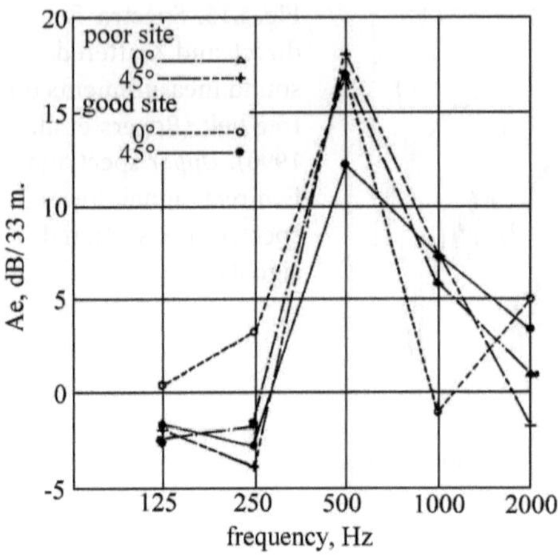

Fig. 4.16. Excess attenuation versus frequency (Leonard and Herrington 1971)

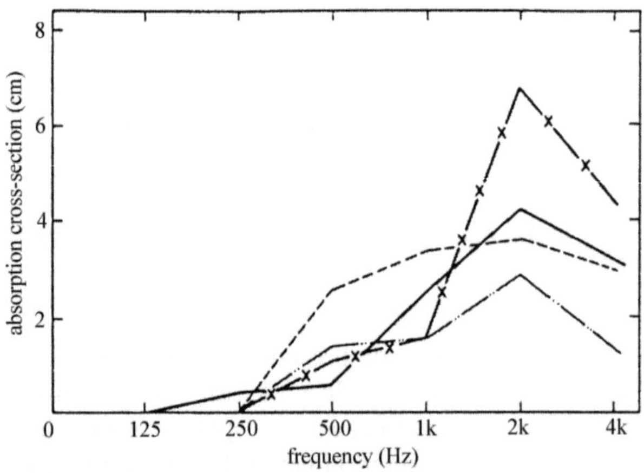

Fig. 4.17. Normalized absorption cross-section versus frequency for pine, maple, ash, magnolia young trees, 2 m height (Bullen and Fricke 1982). Reprinted with permission from Elsevier, copyright 2005

section (cm) versus frequency is shown in Fig. 4.17. At 2 kHz frequency, all species have a maximum absorption cross-section (e.g. for ash 6.5 cm, for pine 4.5 cm, for maple 2.8 cm).

Kellomäki et al. (1976) performed systematic measurements to study the influence of dendrological characteristics of trees on excess attenuation in several stands such as: pine, spruce and mixed (20% birch and other broadleaved species). The density of stems/ha was between 500 and 3,000. The sound source was placed at 12 m from the edge of the stand, reproducing a real situation. The attenuation coefficient was measured as a function of several parameters, such as the percentage of dominant trees, density of trees/ha, total stem number of dominating and dominated trees, basal area, volume, height, age of the stand (Fig. 4.18); and regression equations were calculated. Several of them are given below:

1. Attenuation coefficient P and percentage of dominant trees (p):
   $P = 0.696 - 0.0005, (p)\ r = -0.554$ (significant at 1%, variance explained 31%).
2. Attenuation coefficient P and percentage of self-pruned stems (p_{P}):
   $P = -0.144 + 0.192, (p_{\mathrm{P}})\ r = 0.609$ (significant at 1%, variance explained 37%).
3. Attenuation coefficient P and total stem number (log):
   $P = 0.903 - 0.064, (\log) r = -0.653$ (significant at 1%, variance explained 43%).
4. Attenuation coefficient P and basal area:
   $P = 0.705 - 0.011$, (basal area) $r = -0.493$ (significant at 1%, variance explained 24%).
5. Attenuation coefficient P and height:
   $P = 0.389 + 0.003$, (height) $r = 0.535$ (significant at 1%, variance explained 30%).
6. Attenuation coefficient P and volume:
   $P = 0.615 - 0.0006$, (volume) $r = -0.291$ (significant at 5%, variance explained 8%).

The most important parameters able to explain the variance between 43% and 24% are: the total number of stems, percentage of dominant trees, percentage of self-pruned stems and the height. The volume which represents the major share of the biomass of the stand seems not to be an important parameter in sound attenuation in a stand.

The attenuation as a function of the density and height of trees is given in Table 4.3 for a spruce stand and Table 4.4 for a pine stand. In both cases, the attenuation increases with the increasing height of the trees.

Pal et al. (2000) performed measurements in stands planted around coal mines in India in order to protect the urban area from pollutants and noise. They derived linear regression relationships between the excess attenuation and tree density, average height, canopy branch, trunk diameter, vertical and horizontal light penetration. It was noted that light penetration, which depends on the leaf size, shape and density, is the most explicative parameter, while the density (number of trees/ha) has a<SNM>negligible effect. It was argued that sound waves propagate through the gaps between the trees, even with the maximum plantation density.

In this context of sound absorbers in the forest, it is natural to consider another important constituent of the trees, the bark. The acoustic properties of the bark of different species were studied by Reethof et al. (1977). The absorption coefficient (around 10%) was measured with a standard impedance tube. Small variations were observed between species. The quantitative contribution of the bark to the global behavior of a tree in the acoustical field is not known,

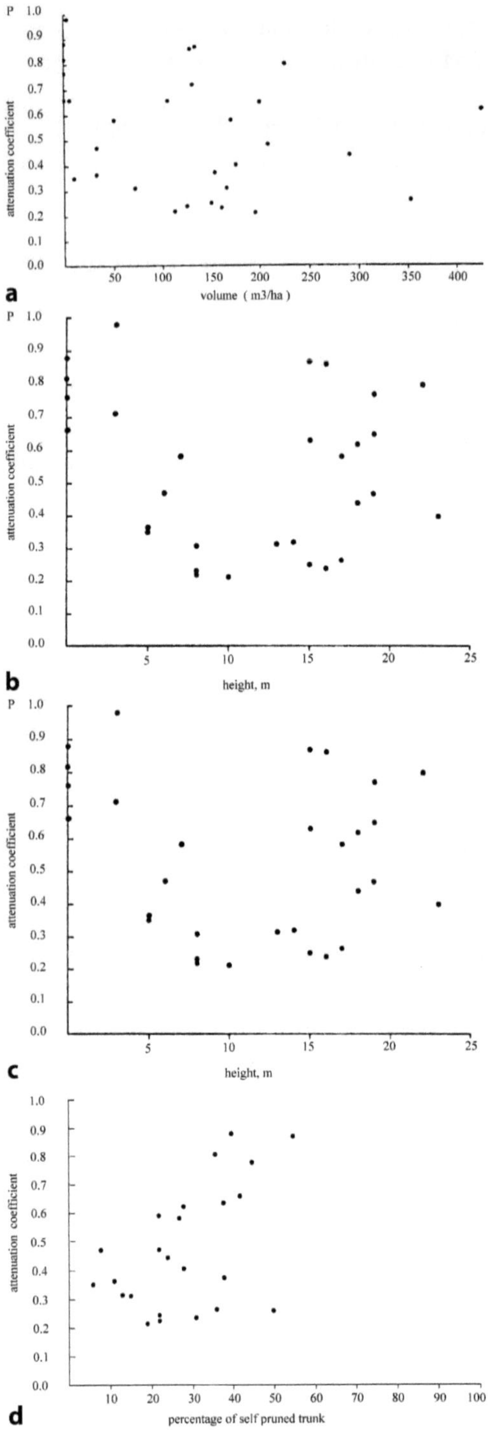

Fig. 4.18. Attenuation coefficient in spruce stand versus (Kellomäki et al. 1976) **a** Volume (m^3/ha), **b** height, **c** age of the stand, **d** percentage of dominant trees

Table 4.3. Attenuation coefficient as a function of density and height of a spruce stand (Kellomäki et al. 1976)

Density	Mean height of the trees (m)												
(stems/ha)	2	4	6	8	10	12	14	16	18	20	22	24	26
	Attenuation coefficient ($\times 10^{-3}$)												
250	715	643	587	547	523	515	523	547	587	643	715	803	907
500	638	566	510	470	446	438	446	470	510	566	638	726	830
750	594	522	466	426	402	394	402	426	466	522	594	682	
1,000	562	490	434	394	370	362	370	394	434	490	562		
1,250	538	466	410	370	346	338	346	370	410	466			
1,500	518	446	390	350	326	318	326	350	390				
1,750	501	429	373	333	309	301	309	333					
2,000	486	414	358	318	294	286	294	318					
2,250	473	401	345	305	281	273	281						
2,750	451	379	323	283	259	251							
3,000	441	369	313	273	249	241							

The attenuation coefficient P_i of the stand i is calculated as: $P_i = V_{ij} \times x_{ij}^2$. V_{ij} is the attenuation at the distance j in the stand i, and X_{ij} is the distance from the source to the point j in the stand i

Table 4.4. Attenuation coefficient as a function of density and height of a pine stand (Kellomäki et al. 1976)

Density	Mean height of the trees (m)												
(stems/ha)	2	4	6	8	10	12	14	16	18	20	22	24	26
	Attenuation coefficient ($\times 10^{-3}$)												
250	961	889	833	793	769	761	769	793	833	889	961	1,049	1,153
500	884	812	756	716	692	684	692	716	756	812	884	972	1,076
750	840	768	712	672	648	640	648	672	712	768	840	928	
1,000	808	736	680	640	616	608	616	640	680	736	808		
1,250	784	712	636	616	592	584	592	616	656	712			
1,500	764	692	636	596	572	564	572	596	636				
1,750	747	675	619	579	555	547	555	579					
2,000	732	660	604	564	540	532	540						
2,250	719	647	591	551	527	519							
2,500	707	635	579	539	515								
2,750	697	625	569	529									
3,000	687	615	559										

but it can be supposed that multiple scattering phenomena are influenced by the acoustical characteristics of the bark.

Modelling sound propagation in a forest is a challenging task (Attenborough 1985; Price et al. 1988; Attenborough et al. 1995). Application of scattering or diffusion theories (Embelton 1966; Barrière and Gabillet 1999; Salomons 2001;

Defrance et al. 2002; van Renterghem et al. 2002; Heimann 2003) has shown that some effects can be predicted. Defrance et al. (2002) found that, close to the trees, meteorological effects enhance the attenuation by the forest belt.

As a conclusion, it can be said that attenuation by the trunks of trees increases linearly with increasing trunk diameter. The increase in attenuation is weaker for higher tree densities. Trees can improve the efficacy of the barriers in downwind conditions. Trunk scattering diminishes the ground effect by reducing the coherence. The ground effect is important below 1,000 Hz. Above this frequency, attenuation by foliage is dominant. Forest reduces the vertical wind and temperature gradient and increases the acoustical efficiency in the case of favorable propagation conditions.

4.3.2 Scattering by Canopy and Foliage

The canopy, composed from branches and leaves, interferes with acoustic energy by scattering, refracting, reflecting and diffracting the acoustic waves. All these phenomena depend on sound wavelength and the dimensions of the boundary between zones of different characteristic impedance (Mueller and Kuk 2000). Smaller objects (like leaves) produce little refraction, diffraction or scattering of waves, but if the objects are numerous, the acoustical properties of the medium might be changed. The waves arriving at any point of the soundfield, after travelling by different paths depending on the geometrical complexity of surfaces and reflections, interfere constructively or not with the soundfield. Such interference between the incident and reflected waves establishes a spatial pattern which is typical for each site. This pattern depends on the speed of sound into the medium. Forest medium is a stratified medium, for which different surfaces act as wave guides, for example the space between the ground and the canopy (neither the ground, nor the canopy are totally absorbing). Temperature and wind gradients might also reflect sound waves, contributing to the formation of guides for sound propagation. The waveguide effect under the canopy or between strata of vegetation might explain the negative "excess attenuations" for certain frequencies or, in other words, this means that sound intensity decreases less than in proportion to the inverse square of the range. An effect of shadow zone for sound propagation during warm sunny days has been also observed.

As noted by Price et al. (1988), the absorption by foliage has been modelled as viscous plate drag, as the result of scattering absorption cross-section or as resonant absorption. Fricke (1984) noted that scattering rather absorption is the more important attenuating phenomenon in the midfrequency (around 1 kHz), while absorption becomes more dominant in the high frequencies. The effect of foliage is well illustrated in Fig. 4.19, in which two maximal attenuations are observed for summer time, the first maximum at about 200 Hz, corresponding to the ground effect, and the second in a high frequency at about 10 kHz, corresponding to the canopy, branches and foliage. The presence of more undergrowth and foliage in summer time explains the rapid increase in attenuation on frequencies higher than 1 kHz. Morton (1975) noted that, close to the ground, a "sound window" can appear which facilitates animal communication in forest (Fig. 4.20).

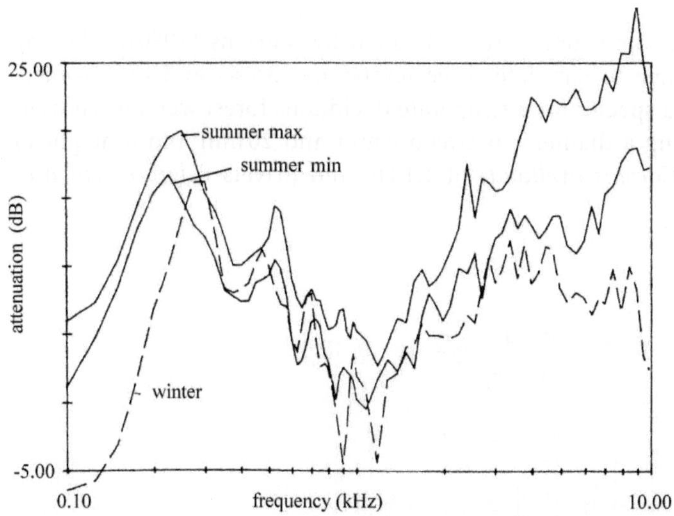

Fig. 4.19. Attenuation versus frequency in a forest stand composed from spruce and oak in alternating bands. Measurements in horizontal range of 72 m, receiver height at 1.2 m (Price et al. 1988). Reprinted with permission from the Acoustical Society of America, copyright 2005. The *continuous lines* correspond to summer measurements – maximum and minimum values; the *broken line* corresponds to winter measurements

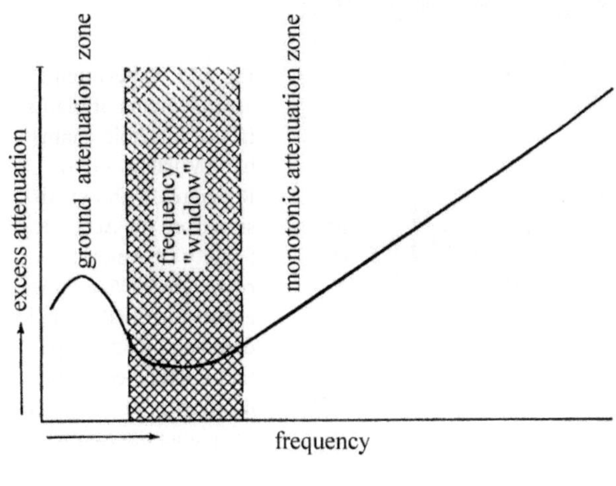

Fig. 4.20. Morton's frequency "window" as an idealized diagram showing the ground attenuation and the monotonic increase in attenuation (Marten et al. 1977)

To obtain a good understanding of the influence of the canopy on sound propagation and transmission through trees, it will be useful to investigate this phenomenon in two typical situations: first under well controlled acoustical conditions (anechoic and reverberant rooms) and second in situ, under natural conditions, for outdoor noise propagation.

4.3.2.1
Measurements in Anechoic Room

Studies in an anechoic room were performed by Martens (1980), selecting different species able to simulate three temperate forests and one tropical forest. The selected species for a temperate deciduous forest were: birch trees (*Betula* spp), having a diameter between 8 mm and 20 mm and a height of 2 m, hazel trees (*Corylus avellana*) of 1.10 m and privets (*Lignum vulgare*)

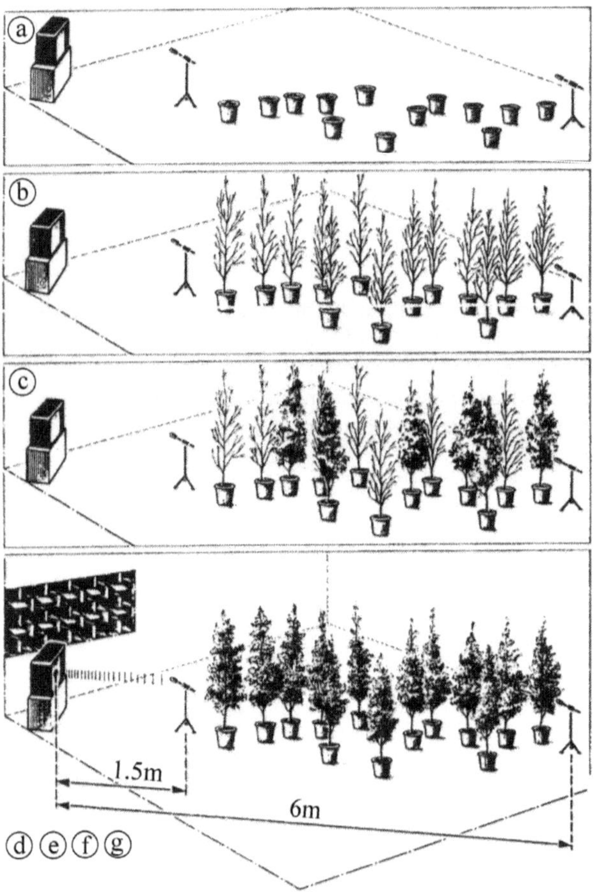

Fig. 4.21. Experimental arrangements of plants in the anechoic chamber (Martens 1980). Reprinted with permission from the Acoustical Society of America, copyright 2005. **a** Flowerpots filled with soil, **b** defoliated plants, **c** partially defoliated plants, **d-g** flowerpots in which the plants were grown

of 0.85 m. The model for the tropical forest was composed of samples from different families (Papilionaceaea, Rubiaceaea, Polygonaceaea, Vitaceaea, etc.) having a height of 2.30 m. The schematic view of the arrangement of plants in the anechoic chamber is given in Fig. 4.21. White noise of 105.5 dB sound pressure level was transmitted in the anechoic chamber, which had a working area of $4.5 \times 4.5\,\text{m}^2$. The total atmospheric absorption over 6 m between the source and the microphone at 10 kHz was 0.5 dB and was neglected. The results are shown in Table 4.5 and in Fig. 4.22. It is notable that the canopy of vegetation has a detectable influence on the noise field, at least in the frequency between 200 Hz and 10 kHz. The spectra in Fig. 4.22a and Fig. 4.22b show the influence of flowerpots filled with soil and root systems only and represent the "ground effect". Up to 8 kHz, no difference was detected. Figure 4.22c and Fig. 4.22g

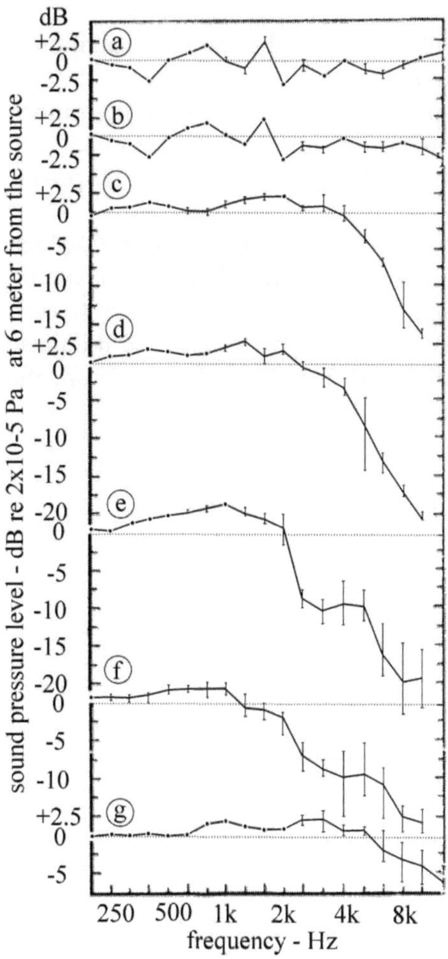

Fig. 4.22. Sound pressure level as a function of frequency for different experimental situations (Martens 1980). Reprinted with permission from the Acoustical Society of America, copyright 2005. **a** Birch trees all sawn down – only earthenware flowerpots with shoots, **b** all birch trees fully defoliated – stems, branches and twigs, **c** 46 birch trees composed of 23 defoliated trees and 23 foliated trees, **d** 46 fully foliated birch trees, **e** 25 fully foliated hazel trees, **f** 26 fully foliated tropical plants of different species, **g** 12 fully foliated privets

Table 4.5. Sound pressure level and attenuation measured in an anechoic chamber with different experimental configurations (Martens 1980). Reprinted with permission from the Acoustical Society of America, copyright 2005

Species	Plants	Total biomass	Total sound pressure level (dB)			Attenu-ation	Frequency drop	Leaf max. size
	(no.)	(kg)	Flower pots only	Foliated plants	Defoliated	Δ SPL	(kHz)	(mm)
Birch	46	8.4	93.0	92.8		−0.2	2–4	70
Tropical	26	11.5	94.7	92.0		−2.7	1.0–1.25	Various
Hazel	25	3.6	95.0	94.1		−0.9	2.0–2.25	130
Privet	12	2.5	96.0	96.0		+1.2	5.0–6.4	40
Birch	46	5.9	93.0	–	93.8	0.0	8–10	20

show the influence of leaves, stems and twigs, and the excess attenuation is detectable, which increased with frequency.

For the experimental conditions reported by Martens (1980), the canopy acts as an amplifier in the midfrequencies (i.e., for birch 200 Hz to 3.2 kHz, for hazel trees 200 Hz to 2 kHz, for tropical plants 200 Hz to 1 kHz, for privets 650 Hz to 5 kHz).

The influence of the biomass is shown in Table 4.5 and in Fig. 4.23, which underline the relationship between the excess attenuation, biomass, maximum dimensions of leaves and the wave number. The specific capacity of each tree species for noise attenuation seems evident. The response to the noise excitation of tropical plants is very different from that of plants from the temperate zone, as can be observed from the data for curves f, g, e.

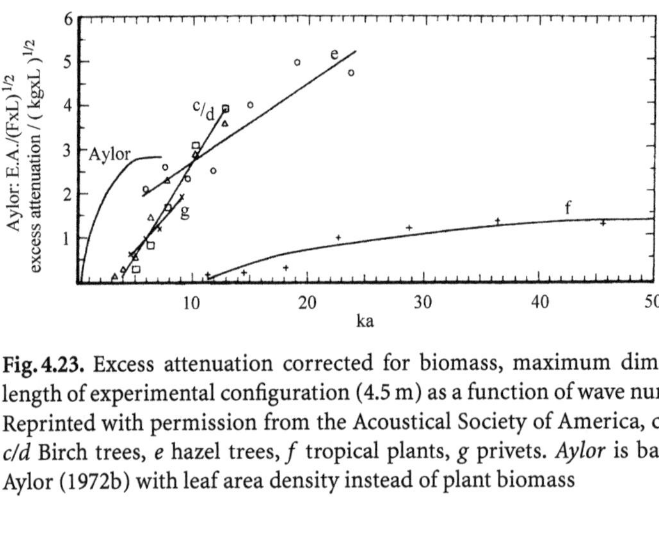

Fig. 4.23. Excess attenuation corrected for biomass, maximum dimension of leaves and length of experimental configuration (4.5 m) as a function of wave number (Martens 1980). Reprinted with permission from the Acoustical Society of America, copyright 2005. *Curve c/d* Birch trees, *e* hazel trees, *f* tropical plants, *g* privets. *Aylor* is based on data given by Aylor (1972b) with leaf area density instead of plant biomass

The efficiency of the filtering action of foliage depends on the noise spectrum and, at the same time, on the sound pressure level of the noise source.

4.3.2.2
Measurements in Reverberation Room

After sound reflection by canopy, branches and leaves, sound refraction takes an important part in the absorption phenomenon. The reverberation room, having a low absorption, is the most appropriate device for detection of the relatively low absorption produced by leaves.

Yamada et al. (1977) proposed a theoretic approach to calculate leaf absorption energy, which depends on several parameters, such as the leaf area, the circular frequency of the leaf, the frequency of the excitation sound, dynamic viscosity and air density.

The absorption coefficient, α_{leaves}, can be calculated with the equation:

$$\alpha_{\text{leaves}} = Gf^2 \tag{4.11}$$

where: f is the frequency, G is a coefficient, depending on the leaf characteristics, (e.g. for a rectangular shape, $G = 0.0002$).

Absorption coefficients and the absorption power of the leaves of trees of different species, versus frequency, were measured in a reverberation room (193 m^3), as can be seen from Fig. 4.24. The influence of an increasing quantity of leaves was expressed by the increasing number of trees in the reverberation room. The absorption coefficient increased with the increasing number of trees in the reverberation room. In this case, the absorption produced by the trees as a whole was measured and it was not possible to differentiate the specific contribution of the leaves. The direct contribution of leaves is shown

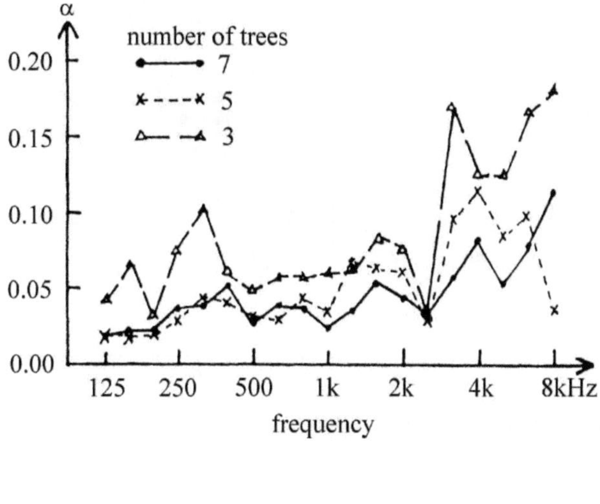

Fig. 4.24. Absorption coefficient versus frequency measured in reverberation room for three, four and five Japanese cypress trees (Yamada et al. 1977)

in Fig. 4.25, in which the absorption power versus frequency is represented for Japanese cypress (trunks with leaves versus trunks only). It can be seen that absorption by trunks only is about one-third of the absorption by trunks

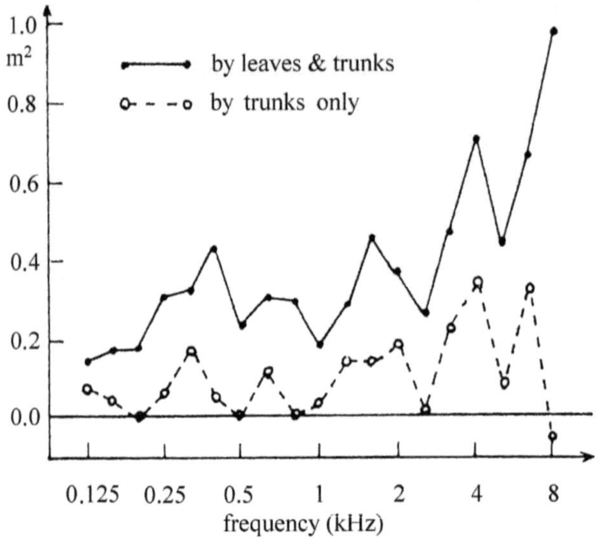

Fig. 4.25. Absorption power versus frequency for Japanese cypress with and without leaves (Yamada et al. 1977)

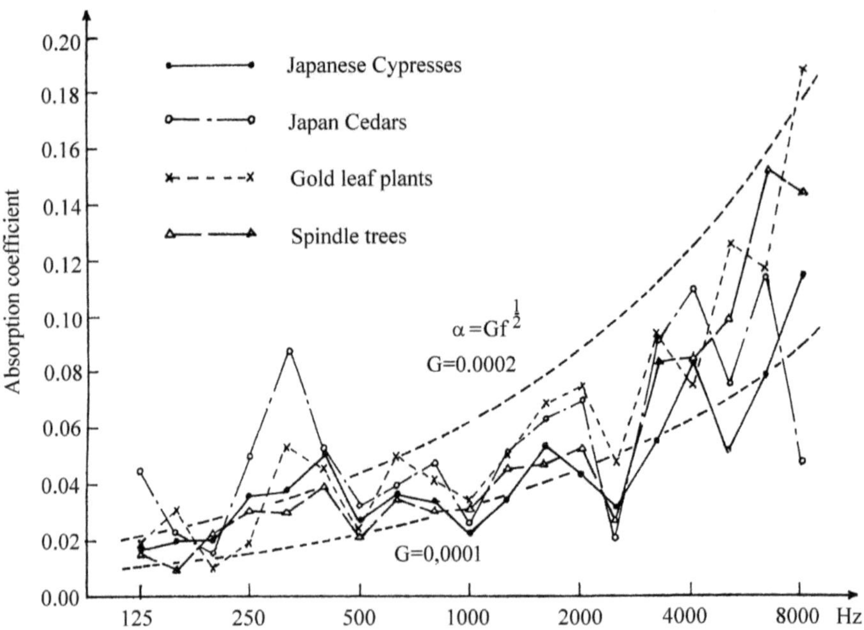

Fig. 4.26. Absorption coefficient versus frequency for different species (Yamada et al. 1977)

with leaves. In this way, it was demonstrated that sound is absorbed by leaves. The same conclusion was deduced from Fig. 4.26, in which the absorption coefficient for different species is represented versus frequency. The proposed hypothesis for the absorption coefficient being proportional to the square root of frequency is valid.

Burns (1979) was interested to understand the mechanism of thermoviscous absorption by white pine (*Pinus strobus*) needles and, for this purpose, measurements in a reverberant room were the best way to answer this question. Fundamental resonances were observed at 4 Hz for 8.8 cm needles and 49 Hz for 2.3 cm needles. Needle-flexing frequency occurred at 20 Hz, particularly at the end of the needle-growing season, when an important amount of identical needles could provide a resonance effect.

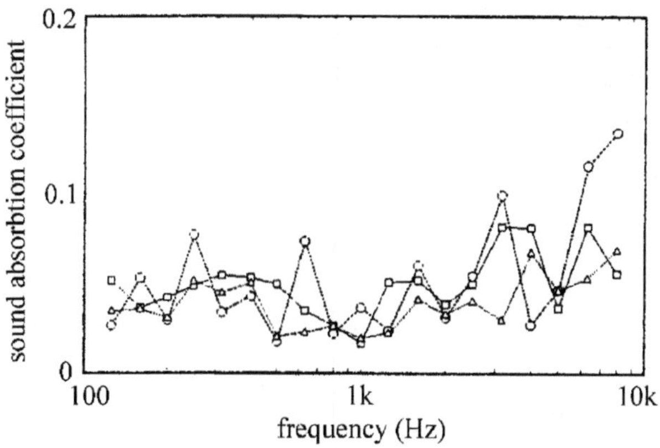

Fig. 4.27. Sound absorption coefficient for sawara cypress with different total leaf surface (Watanabe and Yamada 1996). Reprinted with permission of the Acoustical Society of Japan, copyright 2005. ○ Surface 2.80 m^2, □ surface 4.78 m^2, △ surface 8.70 m^2

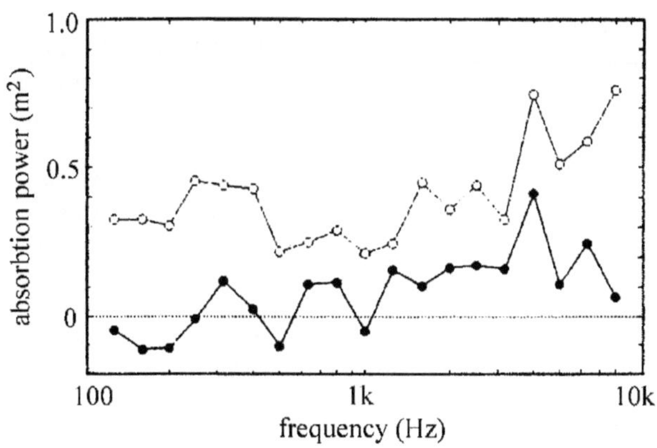

Fig. 4.28. Absorption power versus frequency for sawara cypress with and without leaves (Watanabe and Yamada 1996). ○ With leaves, surface 8.70 m^2, ● without leaves. Reprinted with permission of the Acoustical Society of Japan, copyright 2005

For the measurement of the sound absorption coefficient by foliage, Watanabe and Yamada (1996) introduced several trees (without roots) into a reverberation room and measured the reverberation time with and without trees. One loudspeaker and four microphones were used. The sound source signal was one-third octave band noise, with center frequencies at 125 Hz to 8 kHz.

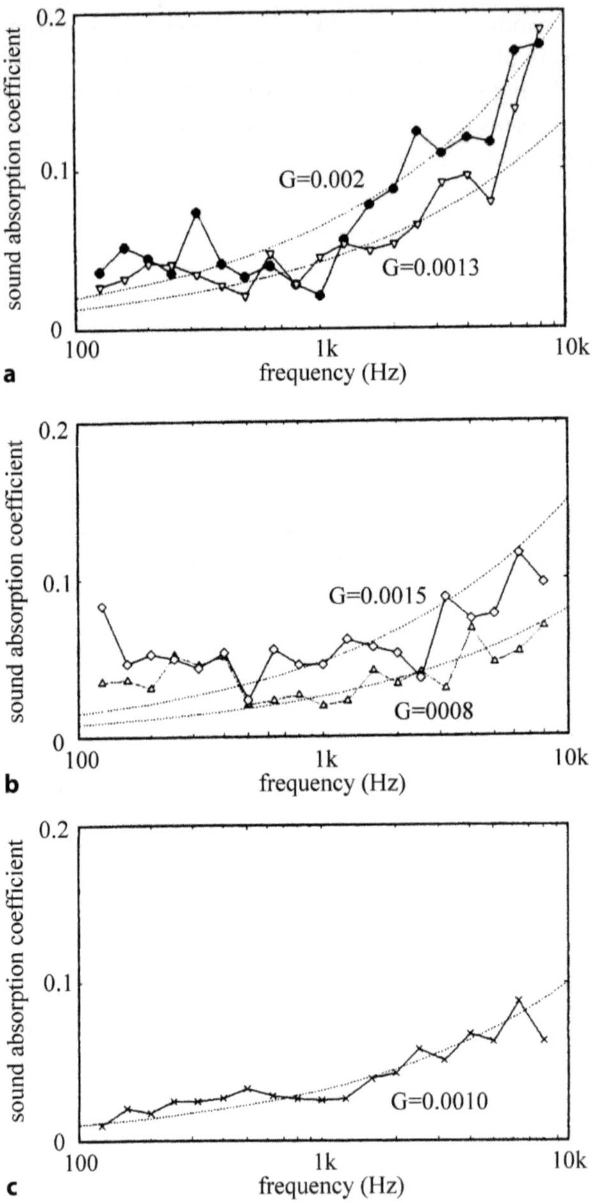

Fig. 4.29. Sound absorption coefficient for trees of different species and different leaves versus frequency (Watanabe and Yamada 1996). Reprinted with permission of the Acoustical Society of Japan, copyright 2005. △ Sawara cypress (*Chamaecyparis pisifera* var. *plumosa*), ● Japanese aucuba (*Aucuba japonica*), ◇ Japanese cedar (*Cryptomeria japonica*), ▽ spindle tree (*Euonymus japonica*)

The absorption coefficient of trees can be calculated as follows:

$$\alpha_m = \frac{55.26xV}{S_m c}\left(\frac{1}{T_m} - \frac{1}{T_0}\right) \tag{4.12}$$

where V is the volume of the reverberation room, S_m is the surface area of the leaves, T_m is the reverberation time with trees and T_0 is the reverberation time of the empty room (without trees).

Figure 4.27 gives the sound absorption coefficient for sawara cypress (*Chamaecyparis pisifera* var *plumosa*) as a function of frequency for three different leaf surfaces, namely 2.80 m^2, 4.78 m^2 and 8.70 m^2. The general trend of curves is very similar. The absorption coefficient is independent of leaf area and frequency.

The calculation of the absorption power – which represents the fraction of the incident acoustic power arriving at the boundary that is not reflected and is therefore regarded as being absorbed by the boundary (Morfey 2001) – and its representation as a function of frequency (Fig. 4.28) has shown that the absorption power of leaves is higher than that of the "skeleton" (composed only of branches). Figure 4.29 represents the variation in sound absorption coefficient versus frequency for different species. The broken lines represent theoretical values determined with the equation $\alpha_{\text{leaves}} = Gf^2$, where G is the frequency absorption factor and f is the frequency. A good agreement is observed between the experimental and theoretical values, for G between 0.0010 and 0.0020.

The attenuation produced by absorption can be calculated with:

$$(Att)_{\text{absorption}} = -10\log\left[1 - \frac{1}{8}GFLf^{\frac{1}{2}}\right] \tag{4.13}$$

where F is the total surface area of leaves per unit volume, $L = 0.3$ m is the diameter of the tree and f is the frequency (i.e. $f = 100$ Hz). Figure 4.30 gives

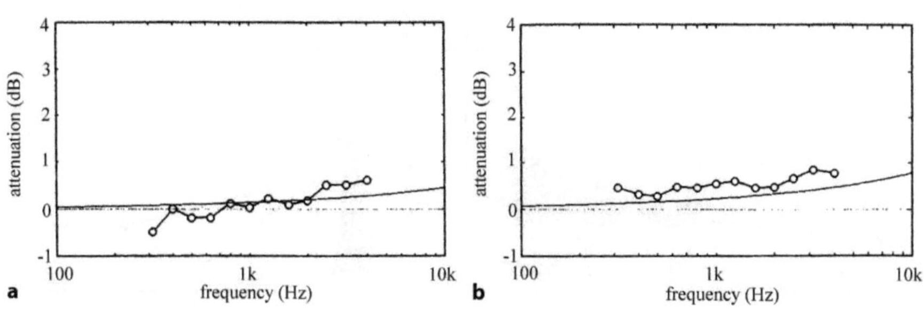

Fig. 4.30. Sound attenuation through absorption by leaves. Theoretical curves and experimental values (Watanabe and Yamada 1996). Reprinted with permission of the Acoustical Society of Japan, copyright 2005. **a** Japanese cedar ($FL = 5.45$, $G = 0.0015$), **b** sudaji ($FL = 6.53$, $G = 0.002$)

theoretical curves and experimental values for sound attenuation through absorption by leaves for two species: coniferous (Japanese cedar, for which $FL = 5.45$ and $G = 0.0015$) and deciduous (Sudajii tree, for which $FL = 6.53$ and $G = 0.002$). In this case, sound attenuation through absorption is less than 1 dB between 300 Hz and 5 kHz.

4.3.2.3 Outdoor Measurements

The mechanism of outdoor sound propagation is summarized in Fig. 4.31. The atmospheric absorption, the ground, the belt of trees, the wind and temperature gradients attenuate the sound which propagates from the source to the receiver along a specific path. Depending on its nature (soft or hard), ground reflections interfere with incident sound, producing attenuation or amplification. The belt of tree acts as a sound barrier. Because of scattering, the canopy of the trees can modify the effectiveness of sound barriers. Wind and temperature vertical gradients refract the sound path (up or down), producing sound shadow zones, contributing or not to the effectiveness of sound barriers (Anderson and Kruze 1992).

In situ full-scale measurements related to the effect of canopy were reported to our knowledge by Lyon et al. (1977), Martens (1981), Piercy and Daigle (1998) and ISO 9613-2 (1996).

Piercy and Daigle (1998) noted that attenuation due to propagation through the canopy increases linearly with the propagation distance, when the radius of a curved ray path is about 5 km. As cited by ISO 9613-2 (1996) when the propagation distance is 10–20 m, the attenuation varies between 1 dB for 250 Hz nominal midfrequency and 3 dB for 8,000 Hz; and, when the propagation dis-

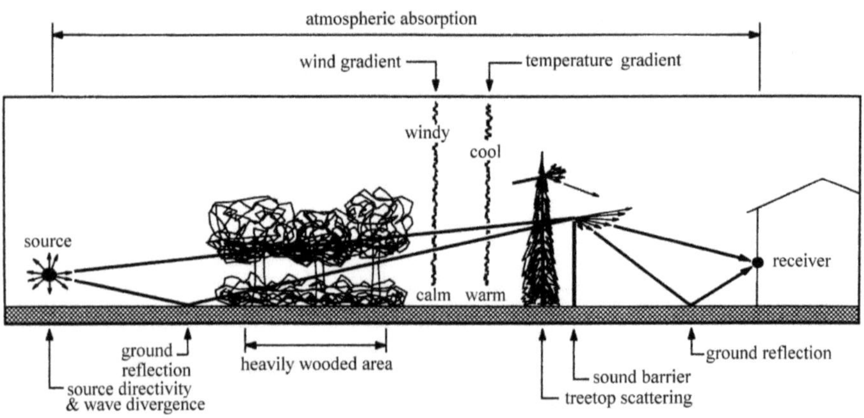

Fig. 4.31. Outdoor sound propagation (Anderson and Kruze 1992)

tance is 20 – 200 m, for the same midfrequencies, the attenuation is respectively 0.04 dB and 0.12 dB.

Excess attenuation as a function of frequency has been studied for sound propagation through and under canopies, for two situations – with and without leaves (Fig. 4.32). The distance between the source and the receiver was 43 m. The frequency spectra with and without leaves diverge above 2.5 kHz. At this frequency, the wavelength is about 12.7 cm and corresponds to the breadth of the leaves. Measurements at various heights in the canopy have shown the leaf effect at 2 kHz.

In situ measurements on foliage attenuation have been reported for tropical forests, temperate forests and for monocotyledonous plants. Eyring's (1946)

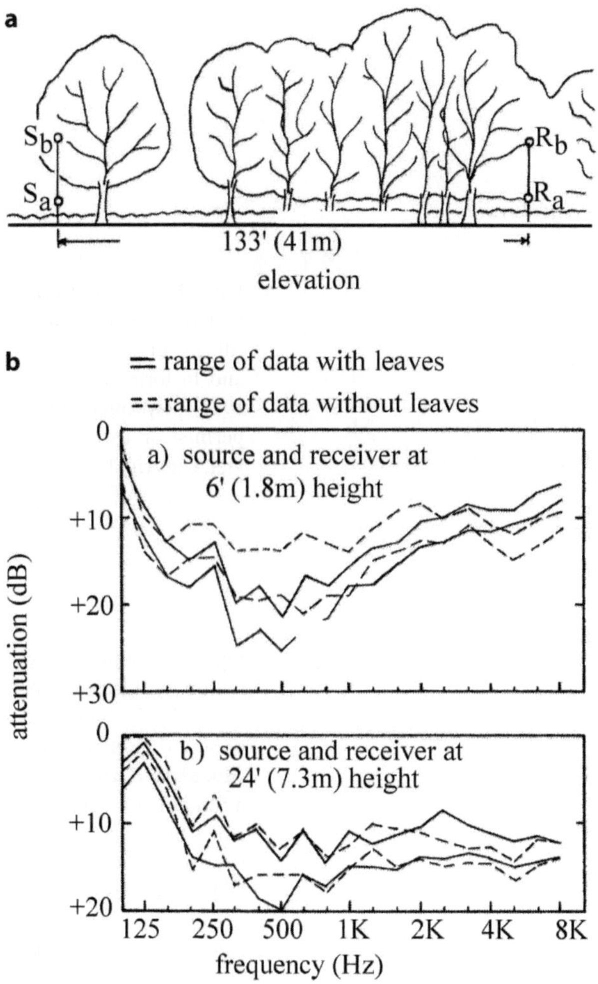

Fig. 4.32. Excess attenuation measured through and under the canopy of maple trees as a function of frequency (Lyon et al. 1977). **a** Under the canopy, **b** through the canopy

pioneering research in tropical rain forests found that foliage has an important role in attenuation, which is correlated with visibility inside the tropical forest. Embelton (1963) noted that, between 200 Hz and 2,000 Hz, vegetation attenuated sound independently of frequency. Branches were identified as resonant absorbers for sound waves between 250 Hz and 1,000 Hz. For deciduous and coniferous trees from temperate zones (cedars, pine, spruce, poplar, elm, maple), no correlation was found between attenuation and visibility. This seems reasonable when comparing tropical and temperate forests. Aylor (1972a, b) studied the influence on the sound field of the canopy of the perennial reed and corn in situ and found an increase in excess attenuation with increasing frequency. Fricke (1984) compared measurements in open ground

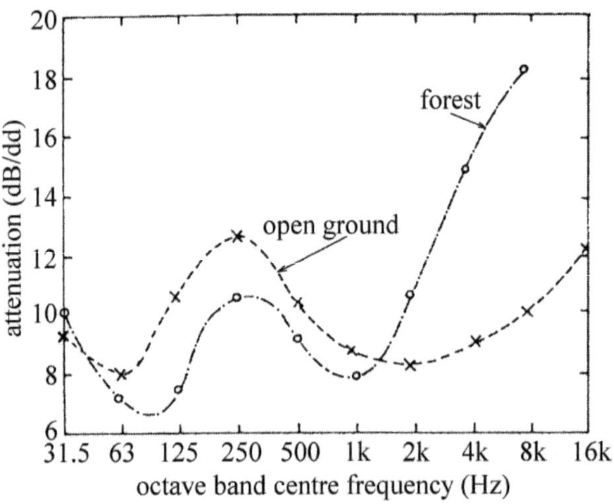

Fig. 4.33. Comparison between attenuation (dB/m) in open ground and in forest (Fricke 1983). Reprinted with permission from Elsevier, copyright 2005

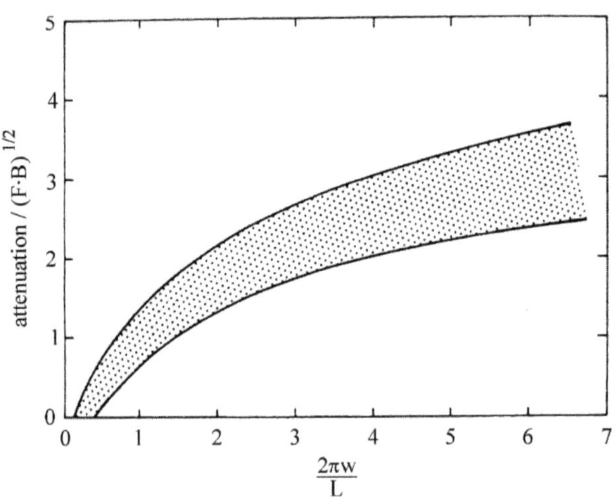

Fig. 4.34. Sound attenuation with leaves (Aylor 1977). Estimation of sound attenuation (dB; *shaded area*) as a function of leaf width corrected with sound wavelength, where *F* is leaf area per unit volume of canopy, *B* is breadth of canopy (m) and *L* is the sound wavelength (m)

and in forest (Fig. 4.33) and noted that scattering by trunks and branches and absorption by foliage and bark are well visible over all frequencies.

Aylor (1977) proposed an empiric approach to quantify noise reduction (expressed as the ratio between attenuation and leaf area per unit volume of canopy and breadth of canopy) and leaf width corrected with the sound wavelength (Fig. 4.34).

To investigate the modes of leaf vibration, Martens and Michelsen (1981) used a laser – Doppler vibrometer system. Laser vibrometry is a suitable technique for measuring vibration velocity in small areas of leaves in a wide range of frequencies (0–100 kHz) without mechanical loading of specimens. The lower limit of the amplitude detection is 1 nm. In a soundfield of 100 dB sound pressure level (*re* 20 µPa), vibration velocities were measured between 10^{-5} m/s and 3×10^{-4} m/s and it was demonstrated that the leaves behave as linear systems. The vibration velocity of leaves are 1–3 orders of magnitude smaller then the vibration velocity of air particles (5×10^{-3} m/s). The leaves behave like plates with different modes of vibration (Fig. 4.35). Only a part

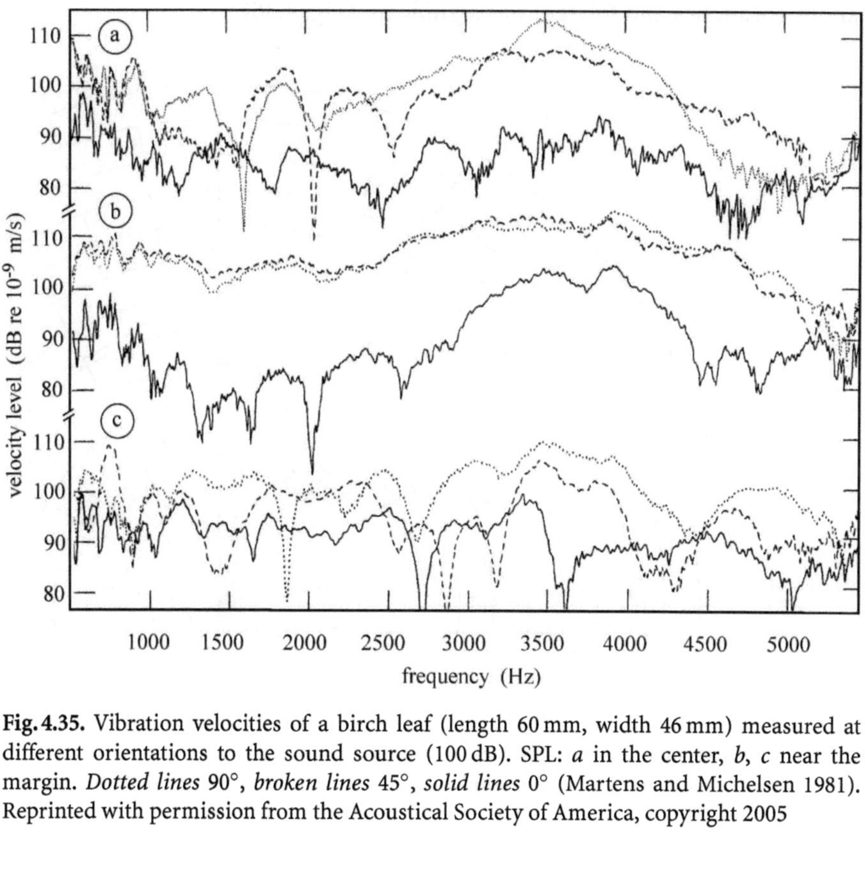

Fig. 4.35. Vibration velocities of a birch leaf (length 60 mm, width 46 mm) measured at different orientations to the sound source (100 dB). SPL: *a* in the center, *b*, *c* near the margin. *Dotted lines* 90°, *broken lines* 45°, *solid lines* 0° (Martens and Michelsen 1981). Reprinted with permission from the Acoustical Society of America, copyright 2005

of the sound energy reaching the leaves will produce vibration energy, the other part will be reflected and diffracted. If the absorption of sound energy is the phenomenon of major importance, the excess attenuation should be linear with the pathlength and foliage density; and this could explain the noise attenuation of plants in the environment.

Aylor (1972a, b) noted that the noise reduction increases with increasing leaf width. Note also that Aylor used in his experiments only monocotyledonous plants, for which the dimensions of the leaves are characterized by a very high length/width ratio. Monocotyledonous (palm with and palm without foliage) behavior was also studied by Bullen and Fricke (1982). The characteristics of the trees studied are the following: the first palm had a trunk diameter of 50 cm and no foliage below 6 m from the ground; the second palm was of about the same size and had very thick foliage extending to the ground and out from the trunk to a radius of 2 m; and the third tree was a big fig. At 1 kHz, the results for palms were similar and the cross-scattering area was 0.4 m. Measurements on a big fig, having a diameter of 1 m and a big canopy with leaves and both large and small branches spreading at 30 m, showed that the normalized cross-scattering area was 2 m. In this case, the scattering produced by the trunk and canopy (branches and foliage) was significant (Fig. 4.36).

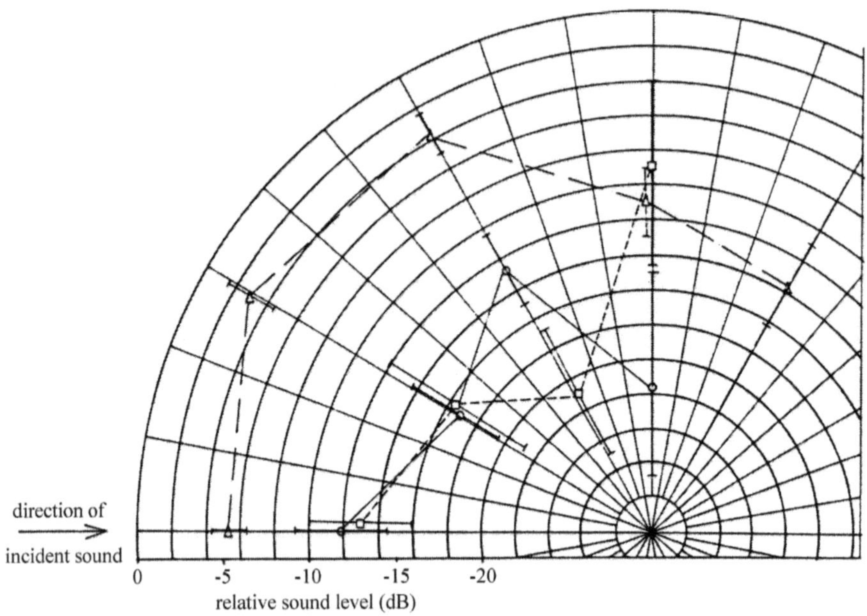

Fig. 4.36. Relative sound level (dB) of sound scattered by trees (Bullen and Fricke 1982). Reprinted with permission from Elsevier, copyright 2005. *Circles* Palm without important foliage, *squares* palm with foliage, *triangles* fig. The direction of the incident sound was between 0° and 90°

The normalized scattering cross-section (σ_S) was measured by emitting a pulse of $\Delta t = 1$ ms duration from the source and recording the level of the reflected pulse. The scattering length is $L_S = 1/(N \times \sigma_S)$, where N is the number of trees/m^2. The absorption length is $L_a = 1/(N \times \sigma_a)$. If the absorption is not a dominant process, we have: $L_a \ll L_S$.

Bullen and Fricke (1982) stated that, at 500 Hz, for a conventional, typical tree of 5 m height and 50 cm diameter, with average foliage having a normalized absorption cross-section of 5 cm, at 2 kHz, the absorption cross-section can be between 10 cm and 15 cm. The variation in the reduced absorption cross-section as a function of frequency for trees with different shape and size of canopy and foliage was shown previously in Fig. 4.17. A peak of absorption cross-section was observed at 2 kHz for all species. The maximum was for magnolia (having large leaves) and the minimum for ash (having relatively small leaves). The wavelengths of these frequencies are compatible with the leaf sizes, since the maximum dimension of the leaves is between 0.5 λ and 1.0 λ. Resonances of branches were identified by Embelton (1963)

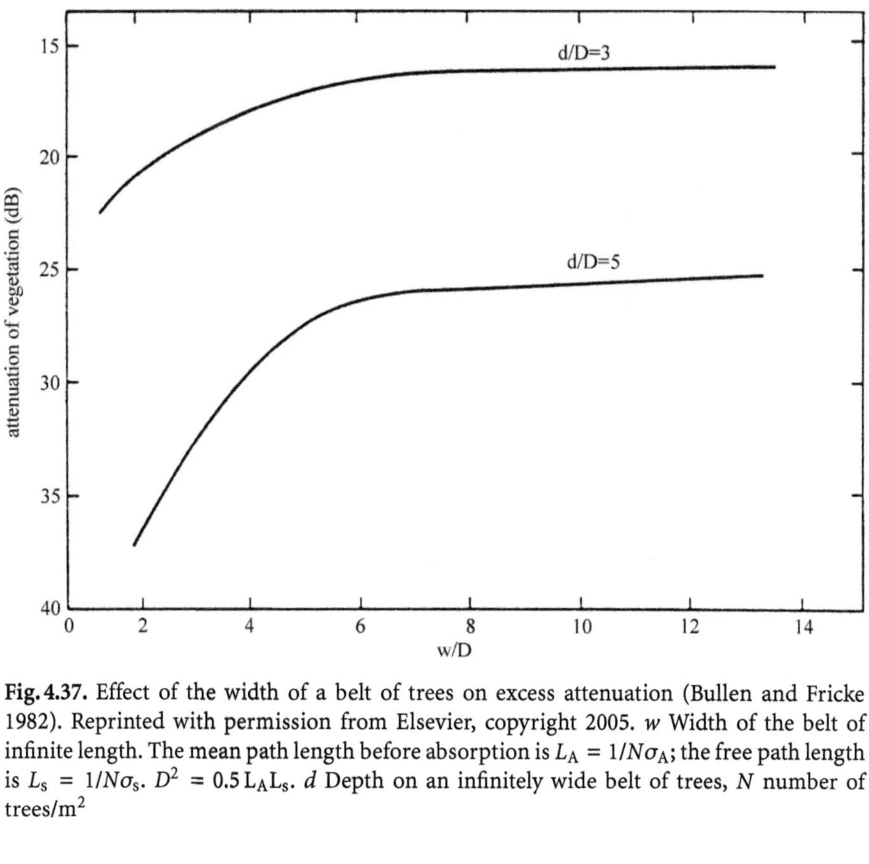

Fig. 4.37. Effect of the width of a belt of trees on excess attenuation (Bullen and Fricke 1982). Reprinted with permission from Elsevier, copyright 2005. w Width of the belt of infinite length. The mean path length before absorption is $L_A = 1/N\sigma_A$; the free path length is $L_S = 1/N\sigma_S$. $D^2 = 0.5\,L_A L_S$. d Depth on an infinitely wide belt of trees, N number of trees/m^2

between 275 Hz and 400 Hz, corresponding to a branch length between 1.2 m and 1.8 m.

The effect of foliage integrated into the width of a belt of vegetation is shown in Fig. 4.37. If the ratio $w/D > 5$, the attenuation is constant (about 26 dB for the ratio $d/D = 5$; and about 16 dB for the ratio $d/D = 3$). For the general case, it was stated that about 3 dB excess attenuation may be gained by making the belt approximately as wide as its depth.

The effect of foliage on excess attenuation can be well observed when comparing experiments in winter and summer (Price et al. 1988). In the absence of leaves, attenuation was considerably low. A peak of attenuation was found at 200 Hz and was attributed to the ground. For frequencies higher then 1,000 Hz, the attenuation gradually increased and was attributed to the trunks and foliage.

The influence of solitary tree foliage density on noise reduction was studied by Schaudinischky et al. (1982). The trees were instrumented as shown in Fig. 4.38a. Measurements "in situ" in a forest stand were performed at different distances from the source, following the scheme presented in Fig. 4.38b. A quality score (Q_{res}) ranging from 10 to 100 was proposed to determine the acoustic

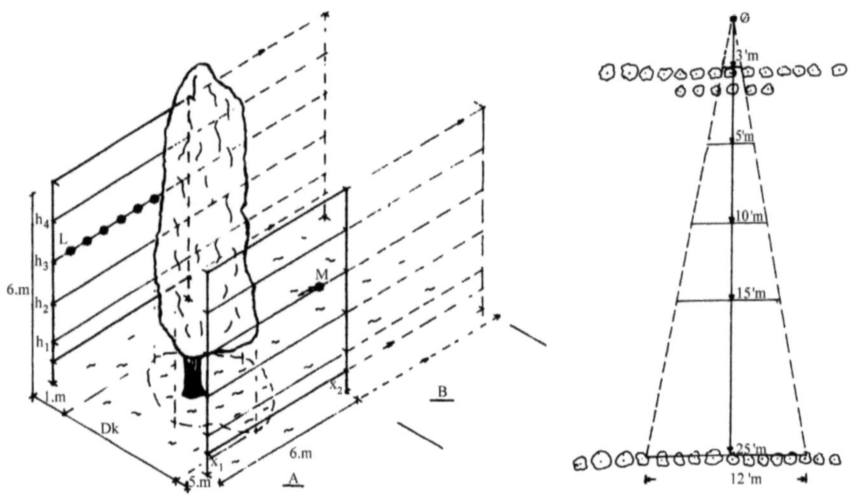

a **b**

Fig. 4.38. Experimental set up for noise reduction measurement of a solitary tree (Schaudinischky et al. 1982). **a** Measurements on a solitary tree. *A* Measurement zone for tree, *B* measurement for free field, *L* noise source – loudspeaker position at different heights (maximum 6 m), *M* motorized mobile microphone, D_k diameter of the canopy, n_1–n_4 measurement points in *vertical direction*, x_1–x_n measurement points in *horizontal directions*. **b** Measurements on forest stands. Only the first two rows of trees are epresented, as well as the last row. The measurements were taken at 3 m, 5 m, 10 m, 15 m and 25 m from the source

quality of each tree, taking into account the foliage density, the growth rate and the noise reduction effectiveness. The measured parameter was the difference in sound pressure level, Δ_L, behind and below the tree, in a frequency range between 63 Hz and 8,000 Hz in one-third octave band analysis.

The sound pressure level as a function of canopy diameter ($\Delta\overline{L}$) was expressed in dB/m and was deduced from the empirical equation:

$$\Delta\overline{L} = \Delta_L / D_k (\text{dB/m}) \qquad (4.14)$$

The quality score was calculated with 4.15:

$$Q_{\text{res}} = \left[60 \log \Delta\overline{L} + 10 \times \mathrm{e}^{0.23} \frac{t_D}{t_{D_{\text{ref}}}} \right] - B \qquad (4.15)$$

where t_D is the growing time (years) to reach 4 m height, $T_{D_{\text{ref}}}$ is the growing time (years) to reach 12 years of age and B is the correction coefficient related to species (0 for coniferous and 5 for deciduous species). Δ_L is the difference in sound level.

The quality score factor for different species is given in Table 4.6. Broadleaves trees score better than pine, whereas cypress, thuya and callitris occupy an intermediate position.

The influence of species on noise reduction as a function of frequency is shown in Fig. 4.39. It is evident that deciduous species (having an important volume of canopy and leaves) are more effective in noise reduction than coniferous species. The influence of the density of leaves and branches (expressed by measurements at different heights in the canopy as a function of frequency) is

Table 4.6. Characteristics of solitary trees (Schaudinischky et al. 1982)

No.	Species	Canopy diameter (D_k; m)	t_D (years)	SPL related to canopy diameter $\Delta\overline{L} = \Delta_L/D_k$ (dB/m)	Quality score factor (Q_{res})
1	*Callitreis verrucosa*	7	8	0.63	55
2	*Thuya orientalis*	5	8	0.74	48
3	*Cupressus sempervirens* var. *Hor.*	6	6	0.78	48
4	*C. sempervirens* var. *pyr.*	4	6	0.58	45
5	*Pinus halepensis*	7	6	0.39	42
6	*P. pinea*	7	10	0.50	46
7	*Acacia cyanophylla*	9	5	0.54	59
8	*Eucalyptus camadulensis*	7	4	0.57	56
9	*Ficus retusa*	9	10	0.69	61
10	*Quercus calliprinos*	7	12	0.53	47
11	*Q. ithaburensis*	5.5	12	0.40	32
12	*Ceratonia siliqua*	7	12	0.54	55

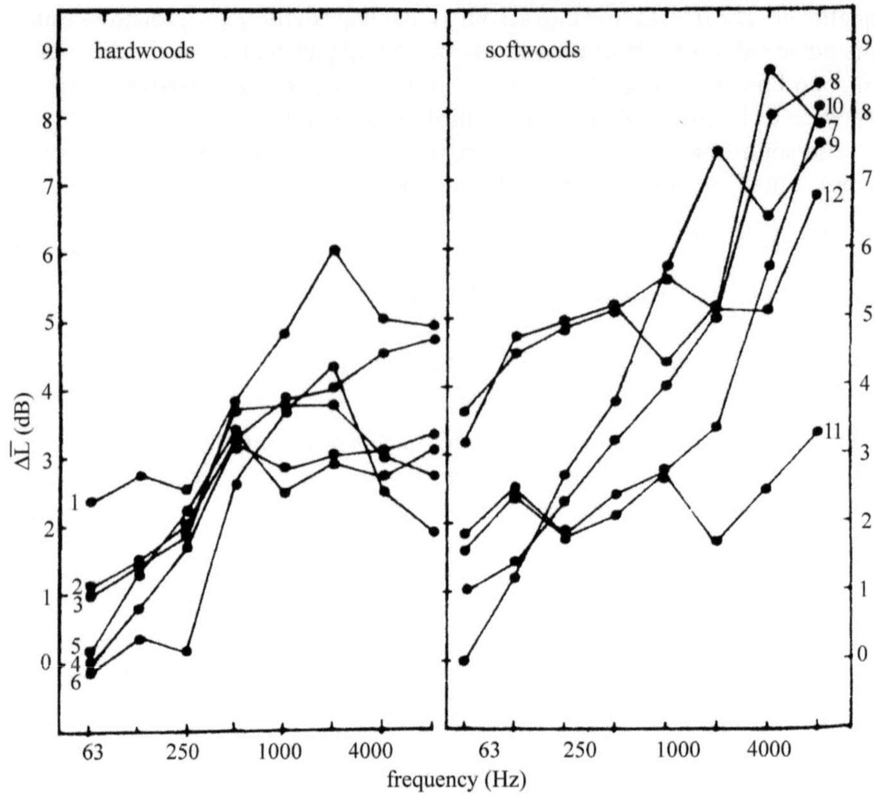

Fig. 4.39. Noise reduction related to canopy diameter, D_k, as a function of frequency and expressed by $\Delta\overline{L} = \Delta_L/D_k$ (dB/m; Schaudinischky et al. 1982). The *numbers* correspond to different species: *1 Eucalyptus camaldulensis, 2 E. camaldulensis, 3 E. camaldulensis, 4 Ceratonia siliqua, 5 C. siliqua* and *Quercus calliprinos, 6 C. siliqua* and *Q. calliprinos, 7 Q. calliprinos, 8 Q. ithaburensis, 9 Collitris verrucosa, 10 Cupressus sempervirens* var. *pyr., 11 C. sempervirens* var. *Hor., 12 Pinus pinea* and understand, *13 P. pinea, 14 P. halepensis, 15 P. halepensis*

shown in Fig. 4.40. The dispersion of measurements in the canopy of *Cupressus* is less important than in *Ceratonia*. The effect of species canopy on attenuation is shown in Fig. 4.41. The dispersion of experimental values is very high and it is difficult to extract a precise law. However, it can be noted that attenuation increases with frequency.

Noise reduction by forest stands composed of the same species was examined (Table 4.7). It was found that broadleaved trees reduce noise better than conifers. Noise abatement is stronger when the foliage extends close to the ground, as in young stands or in the presence of undergrowth. Noise reduction within the stand as a function of the distance from source is shown in Fig. 4.42 for two values of sound pressure level, namely < 0.20 dB and > 0.20 dB. As

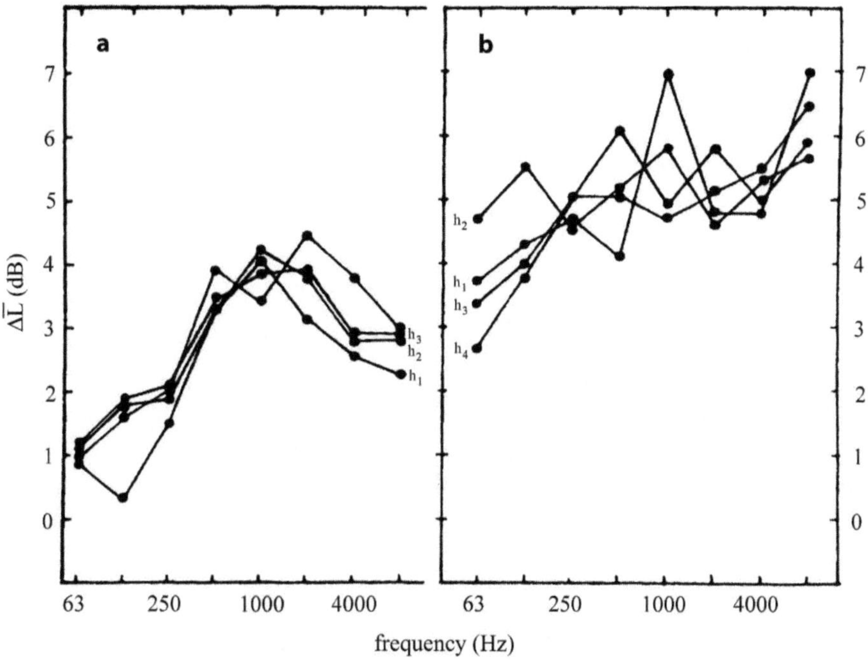

Fig. 4.40. Noise reduction expressed by $\Delta\bar{L}$ produced by solitary trees, as a function of the height in the canopy and frequency (Schaudinischky et al. 1982). **a** *Cupressus*, **b** *Ceratona*

expected, in both cases, the increase in distance from the source has a positive influence on the noise reduction produced by the stand. To achieve a good noise reduction, it was suggested to plant rows which would be felled alternately to maintain dense foliage near the ground. Suitable species with dense foliage are pyramidal *Cypressa*, *Callitris*, *Thuya*, *Ceratonia*, *Eucalyptus*, *Quercus*.

It may be important to distinguish between the effect of a solitary tree, a group of trees and a stand forest (Fricke 1984). The acoustical response of a group of trees is associated with the ground effect. The impedance of the ground over which the sound propagates affects the attenuation rate, mainly in the 250, 500 Hz frequency range. Scattering by the boles and branches and absorption by the bark and foliage are higher-frequency phenomena. These results are similar to those obtained by Kragh (1981). It appears that scattering, rather than absorption as suggested by Aylor (1972a, b), is the more important phenomenon at the midfrequencies. At high frequencies, absorption takes over as the dominant phenomenon. "If scattering is the cause of attenuation through vegetation then more energy is back-scattered and so the decay with time becomes less as the decay with distance increases" (Fricke 1984).

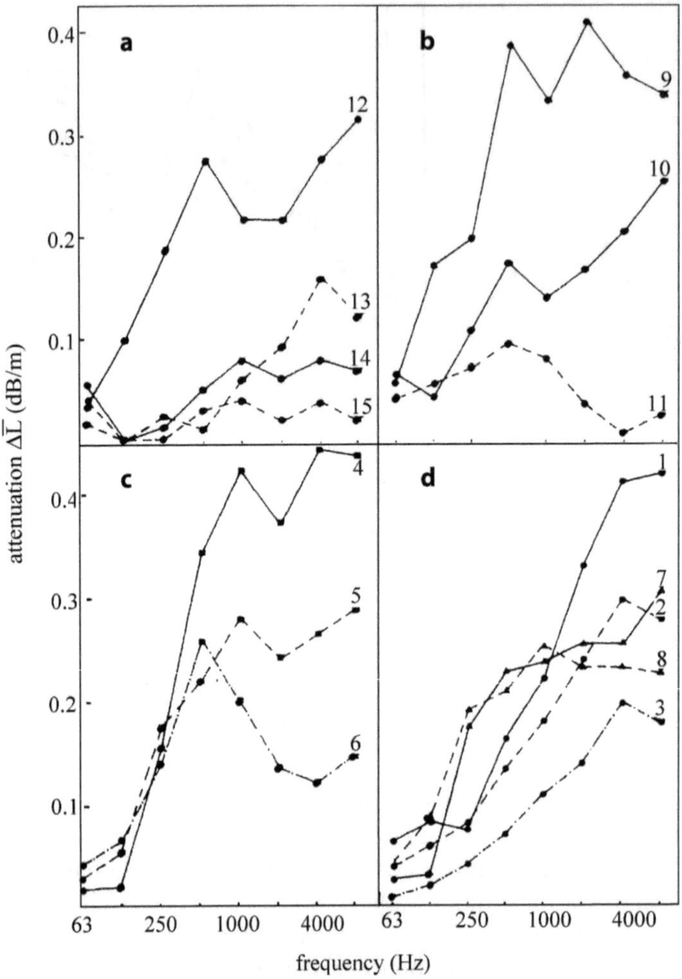

Fig. 4.41. Analysis of noise reduction expressed by $\Delta\overline{L}$ produced by coniferous and deciduous species (Schaudinischky et al. 1982). **a** *Pinus* – measurements on four trees, **b** *Cupressus sempervirens* – measurements on three trees, **c** *Ceratonia siliqua* – measurements on three trees, **d** *Eucalyptus camaldulensis* – measurements on five trees. The *numbers* correspond to different species, as noted in Fig. 4.39

A considerable amount of data (15 stands, Table 4.8) on the "in situ" attenuation rate of forests was published by Tanaka et al. (1979). The profile of stands was complex, composed of coniferous (Japanese black pine, Japanese cedar, Japanese red pine) and deciduous species (mainly beech), as can be seen from Fig. 4.43a. The regression analysis between noise attenuation and distance was established using linear or exponential models ($Y = ab^{\log x}$, where Y is attenuation, x is distance, a and b are experimental coefficients). With the space

Table 4.7. Characteristics of forest stands of different species (Schaudinischky et al. 1982)

No.	Species	Tree density (no./ha)	Average height (m)	Diameter at 1.30m (cm)	Degree (%)	Canopy (m)	SPL related to canopy diameter $\Delta\overline{L}$ [(A)/m; dB]
1	*Eucalyptus camadulensis*	6,000	2.5	1	80	0.20	0.28
2	*E. camadulensis*	3,000	6.0	7	80	0.50	0.18
3	*E. camadulensis*	1,000	14.0	15	50	2.0	0.11
4	*Ceratonia siliqua*	620	5.0	20	80	0.50	0.32
5	*C. siliqua* and *Quercus calliprinos*	700	3.5	10	80	0.20	0.22
6	*C. siliqua* and *Q. calliprinos*	400	3.5	10	40	1.0	0.14
7	*Q. calliprinos*	2,500	3.0	8	90	0.20	0.25
8	*Q. ithaburensis*	2,500	4.0	6	80	0.20	0.24
9	*Callitris verrucosa*	1,000	8.0	15	70	1.0	0.30
10	*C. sempervirens* var. *pyr.*	800	14.0	18	40	0.50	0.16
11	*C. sempervirens* var. *Hor.*	800	14.0	18	60	0.50	0.06
12	*Pinus pinea* – understand	1,500	12.0	15	80	0.20	0.24
13	*P. pinea*	600	12.0	20	70	2.50	0.08
14	*P. halepensis*	1,200	12.0	13	80	2.0	0.06
15	*P. halepensis*	350	18.0	31	50	10.0	0.02

available in this book it has been decided to select only four stands, for which the regressions are shown in Fig. 4.43b, c.

From previous data, it seems that the factors which have a positive influence on the efficiency of forest stands for noise attenuation are: higher stand density, mixed species of trees, larger quantity of leaves (Table 4.9). Measurements in summer and in winter for deciduous stands clearly show the effect of leaves on attenuation.

Here is the place to mention the remarkable pioneering activity (Martens 1981) in the Department of Botany at the University of Nijmegen (The Netherlands) in the field of noise abatement with vegetation and landscape planning. As an example in what follows, the relationships between noise attenuation and four types of forest vegetation (stands, belts, etc.) are described in Tables 4.10, 4.11, 4.12. The acoustic climate was expressed by the variation in excess attenuation versus the frequency in planted forests during 1959–1961, located in Flevopolder and composed of the following species: beech (*Fagus sylvaica*), ash (*Fraxinus excelsior*), spruce (*Picea abies*), poplar with mixed

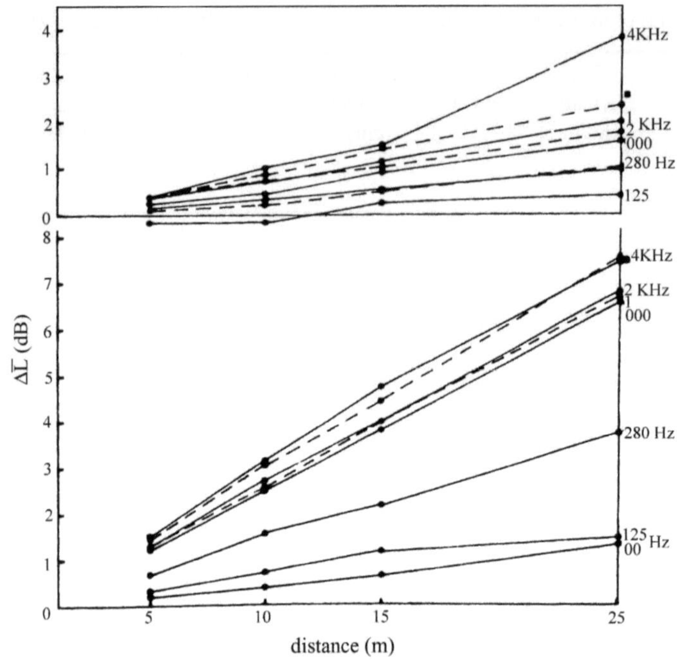

Fig. 4.42. Noise reduction (expressed by $\Delta\bar{L}$) as a function of the distance in the stand, for two levels of excitation (Schaudinischky et al. 1982). The *upper graph* is for L_5 <0.20 dB(A); and the *lower graph* is for L_{50} > 0.20 dB(A)

deciduous species (*Populus x canadensis, Quercus robur, Carpinus betulus, Corylus avellana, Alnus glutinosa, Acer campestre*), or in a belt of spruce (*Picea sitchensis*) and various deciduous trees in the Botanical Garden of the Nijmegen University – the *Stellario carpinetum* (Fig. 4.44). A flat grassfield covered with short grasses of 50 mm maximum height was taken as a reference for the acoustical measurements. The measurements were performed with white noise. The sound pressure level of unfiltered noise was 104 dB and 20 μPa at 6 m in the front of the woofer during all experiments. The microphones were placed at 1.2 m and 3.9 m from the soil; and the distances between the source and the receiver were 6, 12, 24, 48 and 96 m. The frequency spectrum was measured at each distance and analysed in one-third octave bands with center frequencies between 50 Hz and 10 kHz.

From Fig. 4.44a–d, one observes the effect of frequency on excess attenuation for all experiments. A first maximum frequency (noted f') in the low frequency domain (< 250 Hz) can be seen, related probably with the experimental arrangement and is produced by the destructive phase difference between the direct and ground-reflected sound waves (angle φ). For some experiments related to forest stands, f' is between 160 Hz and 250 Hz, while for the grass

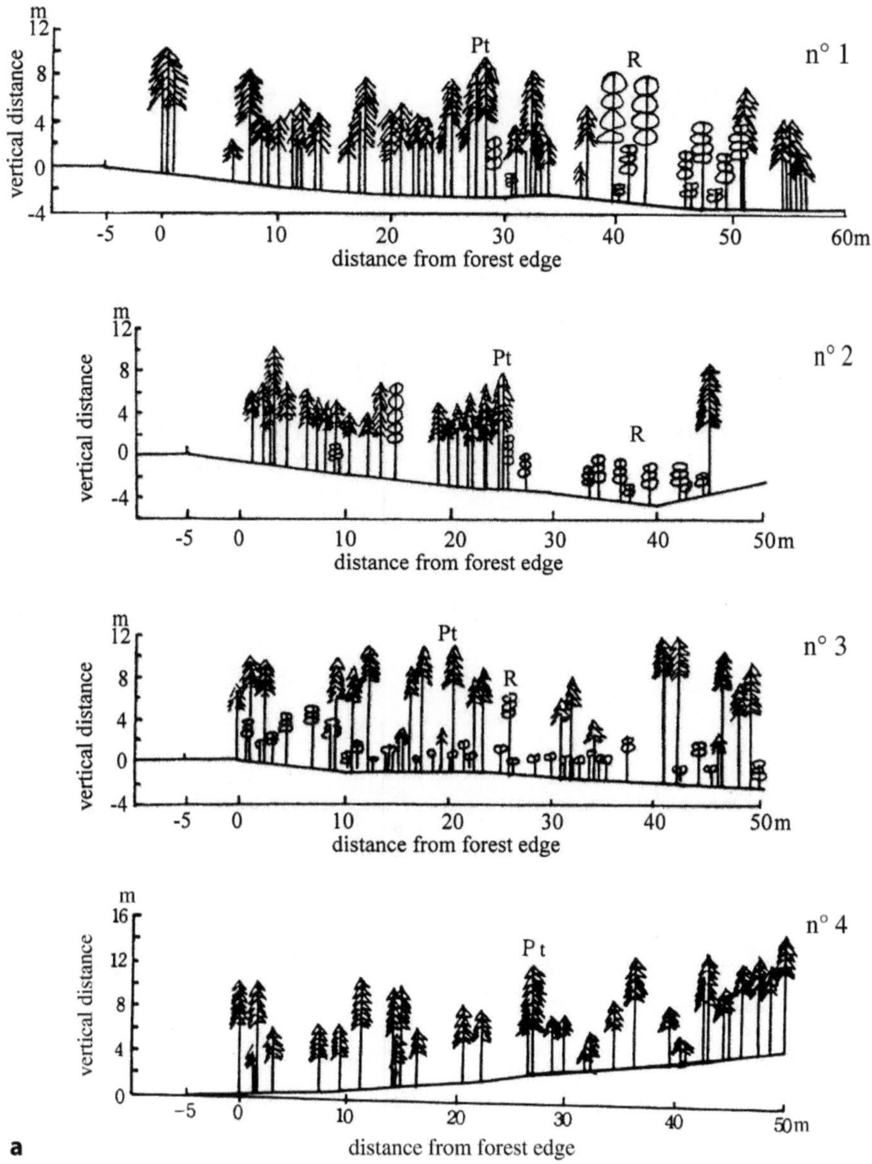

a

Fig. 4.43. Stands for noise attenuation measurements (Tanaka et al. 1979). **a** Profile of stands. *Stand 1* Dominant species – *Pinus thunbergii* (*Pt*), mixed with *Robinia* spp (*R*), density 2,900 trees/ha, 7.2 m height. *Stand 2* The same species as stand 1, density 2,500 trees/ha, 7.8 m height. *Stand 3* The same species, density 1,900 trees/ha, height 9.1 m. *Stand 4 P. thunbergii*, density 3,500 trees/ha, 6.7 m height. **b** Linear regression relationships between noise attenuation [−dB(A)] and distance (m), for the stands represented previously. **c** Logarithmic regression relationships between noise attenuation and distance, for the same stands

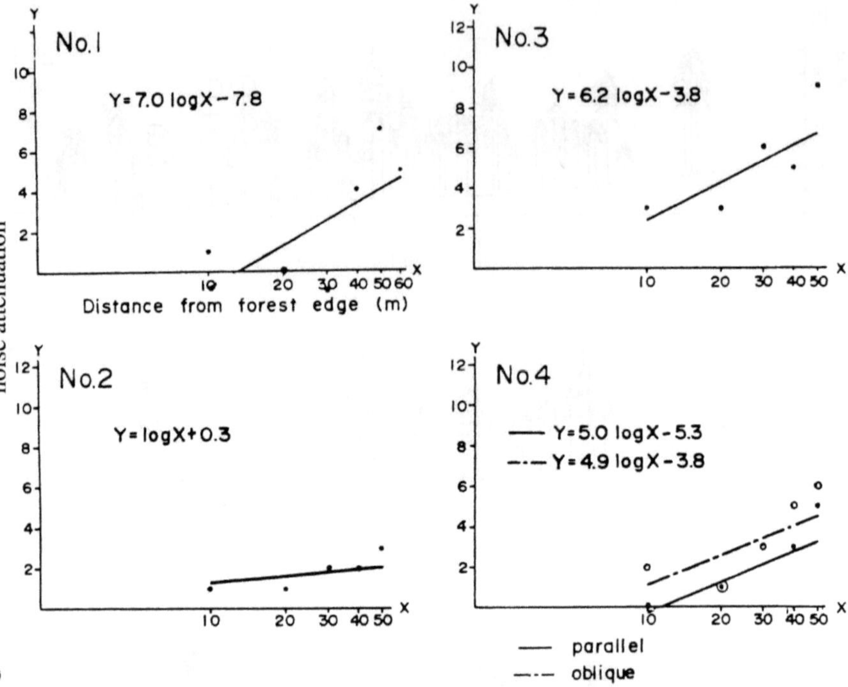

b

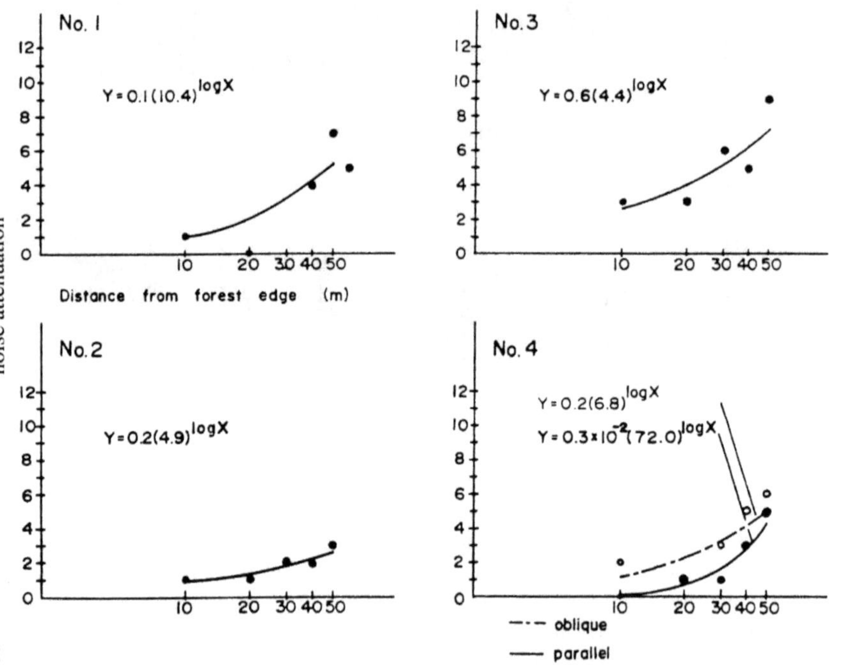

c

Fig. 4.43. (continued)

Table 4.8. Dendrologic characteristics of stands and corresponding attenuation as a function of distance into the stand (Tanaka et al. 1979)

Stand no.	Dominant level				Dominate level						
	Dominant species	Density	Height	Basal area	Density	Relative attenuation (dB) as a function of distance					
		(No./ha)	(m)	(m^2/ha)		10 m	20 m	30 m	40 m	50 m	60 m
1	Pine – mix	2,900	7.2	16.1	Thin	1	0	−1	4	7	5
2	Pine – mix	2,550	7.8	15.9	Thin	1	1	2	3	3	
3	Pine – mix	1,900	9.1	26.0	Thin	3	3	6	5	9	
4	Pine – pure	3,500	6.7	23.8	Thin	2	1	3	5	6	
5	Pine – mix	3,000	13.7	44.2	Thin	0	0	1	1	5	
6	Sugi – pure	4,750	6.0	31.1	Thin	1	2	4	6	8	
7	Sugi	3,400	2.6	21.7	Dense	1	3	3	5	6	
	Hinoki	1,288	–	–		1	4	6	7	9	
8	Sugi – pure	3,920	4.0	45.1	Medium	2	4	5	6	9	10
9	Sugi – pure	4,670	7.3	39.7	Medium	2	4	7			
10	Sugi – pure	4,800	7.7	38.3	Medium	3	4	7	7		
11	*Alnus*	3,000	10.9	35.6	Dense	1	3	6	8	9	
	Robinia	300	4.0	0.4		0	1	5	6	7	
12	*Alnus*	2,600	9.1	26.1	Dense	4	3	7	9	10	
	Robinia	1,400	3.3	1.6		0	1	3	4	6	
13	*Fagus, Lindera*	1,800	7.5	42.8	Dense	3	1	5	5	7	
	Clethera	4,300	1.0	1.5		3	4	7	8	8	
14	*Fagus, Quercus*	1,500	11.0	49.0	Dense	2	4	3	4	6	
	Acer	300	2.5	0.7		1	4	4	6	8	
15	*Fagus, Quercus*	2,500	10.5	67.2	Dense	3	2	5	5	8	
	Clethera	500	1.5	0.5		2	3	5	6	7	

Table 4.9. Best fit reverberation parameters and corresponding prediction of attenuation (Huisman and Attenborough 1991). Reprinted with permission by the Acoustical Society of America, copyright 2005

Parameter	Frequency (Hz)				
One-third octave band center frequency	500	630	1,000	2,000	4,000
Absorption (dB/s)	0	0	0	3	10
Best fit parameters:					
(a) Diameter *De* (m)	0.01	0.02	0.04	0.08	0.16
(b) Reflection factor *Re*	0.1	0.1	0.1	0.1	0.1
Scattering attenuation at 100 m:					
(a) Total scattering attenuation (dB)	−0.8	−1.6	−3.1	−6.3	−12.8
(b) Direct field attenuation (dB)	−0.8	−1.7	−3.3	−6.6	−13.2

Table 4.10. Some characteristics of forests and stands experimented by Martens (1981)

Species	Parameters of trees (m)		Distance (m) between:		Ground	Canopy
	Diameter	Height	Rows	Trees		
Beech forest	11	7.5	2.5	1.35	50 mm litter	Closed
Ash forest	4	6.0	1.4	1.1	Herbs, mosses	Open
Spruce, fir forest	11.5	8	2	1.8	50 cm layer	Closed
Spruce belt	10	7.5	–	–	30 cm grass, reeds	Closed
Mixed poplar forest	29	12	3.3	7.7	Dry wood, mosses	Open
Mixed deciduous species – *Stellario carpinetum*	Various	17	–	–	20–60 mm litter	Open

field it is close to 1,000 Hz. For the spruce belt, f' is between 500 Hz and 800 Hz. These differences could be explained by the different composition of soil surfaces (dry matter, water content, porosity). Forest soil is softer than the soil of the grass field. The second excess attenuation maximum between 2,000 Hz and 4,000 Hz, which is present in all forest types and absent in the grass field, can be attributed to leaves. In the midfrequency range, between 1 kHz and 2 kHz, the excess attenuation is relatively low and quite constant, in deciduous forest stands. This zone, around 2 kHz, was referred to as the "sound window" in the acoustic climate of the plant community and was considered important for the acoustic communications of birds and animals living in the forest. In the high frequency range, a second, but less important maximum can be observed for the measurements in forest only (not in the grass field). It was suggested that this second maximum could be attributed to propagation phenomena in a waveguide (tree branches).

The recapitulation of the experimental data presented by Martens (1981) allows to note that, in most of the one-third octave band studied, in beech and ash forest stands the excess attenuation was at least 10 dB per 100 m with the receiver and source at the same height (1.2 m) and was at least 5 dB per 100 m with the receiver at a height of 3.9 m. Compared with beech and ash stands, in mixed deciduous stands, a *sound window* was detected around 2 kHz, and the ground effect was extended more towards the high frequency range, compared with beech and ash stands. In coniferous spruce stands, the highest excess attenuation was measured, such as 10 dB per 100 m with the receiver at 1.2 m height and 7 dB per 100 m with the receiver at 3.9 m. In the spruce belt, the attenuation, for the same conditions as for the spruce stand, was respectively 7 dB per 100 m and 4 dB per 100 m. As expected, the highest attenuation was found in closed forest stands and not in tree belts.

Martens (1981) concluded that trees, a belt of trees at least 12 m wide and forest stands can be efficiently used to abate noise pollution in urban areas.

Table 4.11. Centre-frequency of one-third octave bands at which appears the first sound pressure minimum (Martens 1981). Reprinted with permission from Elsevier, copyright 2005

Parameter	Distance (h) between source (s) and microphone (r); and path length (Δr) difference between direct and soil surface-reflected sound waves								
	$h_s = h_r = 1.2$ m						$h_s = 1.2$ m; $h_r = 3.9$ m		
	Distance	6 m	12 m	24 m	48 m	96 m	24 m	48 m	96 m
	Δr (m)	0.46	0.24	0.12	0.06	0.03	0.39	0.19	0.10
	Grassfield								
Frequency	Hz	315	500–640	500–800	640–800	500–1,000	400	500	500
Phase	Angle	27	17–53	76–115	130–140	155–166	15	79	126
	Spruce fir belt								
Frequency	Hz	315	250–315	315	315	200–250	250	250	250
Phase	Angle	27	97–115	140	158	173	76	130	155
	Beech forest								
Frequency	Hz	160	250	250	200–250	200–250	200	200	200
Phase	Angle	58	115	148	166	173	97	137	158
	Spruce fir forest								
Frequency	Hz	200	200–250	200	200	200	200	200	200
Phase	Phase	58	115–130	155	166	173	97	137	158
	Stellario carpinetum								
Frequency	Hz	250	250	250	250	–	250	250	–
Phase	Angle	58	115	148	166	–	76	130	–
	Mixed poplar forest								
Frequency	Hz	250	250	250	250–315	250	250	250	250
Phase	Angle	58	115	148	158–166	173	76	130	155
	Ash tree								
Frequency	200	250	250	250	250	200–250	–	200	250
Phase	83	115	148	166	173	137	–	137	155

NB: The phase-reflecting angle in degrees is calculated as $\varphi = \left(\frac{1}{2} - f'\frac{\Delta r}{c}\right) \times 360$, where c is the sound velocity in air and f' is the frequency at which the first maximum occurs.

Table 4.12. Excess attenuation as a function of center-frequency analyzed in one-third octave bands (Martens 1981)

Vegetation	Excess attenuation (dB/100 m, at frequencies ranging over 1 – 10 kHz)									
	1 Hz	1.25 Hz	1.6 Hz	2.0 Hz	2.5 Hz	3.2 Hz	4.0 Hz	5.0 Hz	6.4 Hz	8.0 Hz	10 Hz
Beech tree	12.7	10.4	10.1	7.9	9.7	8.9	11.3	13.3	15.0	16.7	18.5
Ash tree	9.0	8.9	11.7	8.6	13.7	15.6	16.6	18.4	21.7	24.8	28.5
Poplar forest	8.9	8.0	9.4	6.9	8.3	9.1	13.4	16.3	20.5	21.6	24.0
Spruce forest	9.6	8.5	13.5	15.3	19.0	21.5	24.7	28.1	33.2	37.4	35.0
Spruce belt	7.2	7.1	11.0	8.9	11.8	13.7	19.2	21.4	21.3	27.2	29.8

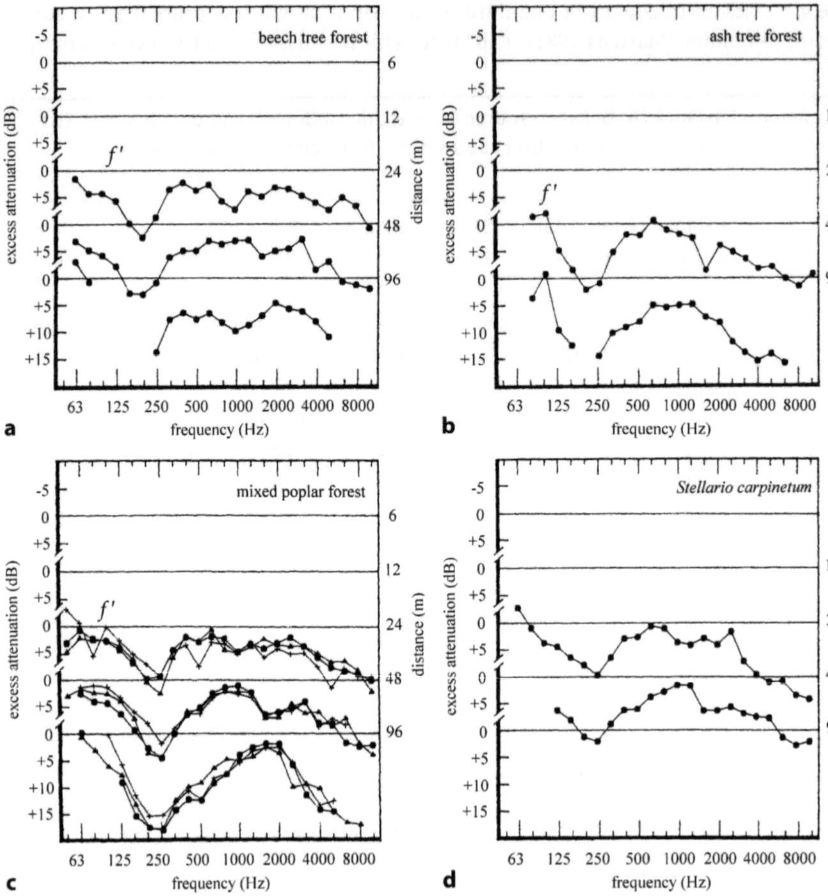

Fig. 4.44. Excess attenuation versus frequency for different distances between source and receiver (Martens 1981). Reprinted with permission from Elsevier, copyright 2005. **a** Beech forest, **b** ash forest, **c** mixed poplar forest, **d** *Stellario carpinetum*, **e** spruce forest, **f** spruce belt, **g** grassfield

4.3.3 Reverberation in a Forest Stand

The well audible reverberation in a forest demonstrates interference between the direct and ground-reflected sound and the scattering effects induced by trees and branches, the ground and possible air turbulence and diurnal variations in meteorological conditions.

It was supposed that the reverberant field consists of randomly reflected particles for which the source location is not relevant. Transient acoustic signals are distorted by the reverberation which is produced by the superimposition

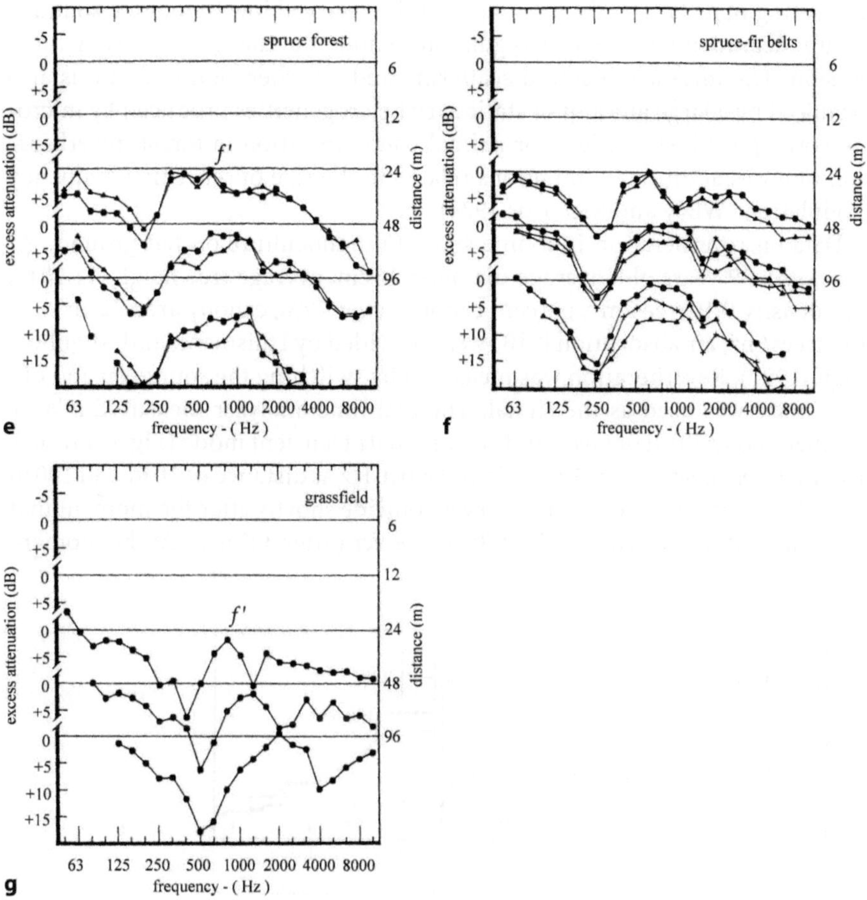

Fig. 4.44. (continued)

at the receiver of direct waves and reflected waves. Because of the complex distribution of reflecting surfaces in the forest, the variety of possible paths for the reflected waves causes an important distortion of transient sound waves.

The factors affecting reverberations are: the directivity of the source and receiver, the carrier frequency and the wavelength of the sound in relation to the dimension and shape of the scattering surfaces. Compared with low frequencies, reverberation increases for frequencies higher than 3 kHz, because of a greater scattering during sound transmission through forest. Reverberation has a masking effect for the long-distance transmission of sound, which can be observed through sharp changes in sound intensity, random amplitude fluctuations accumulating from nonstationary heterogeneities in the propagation medium and signal dispersion produced by numerous reflecting, refracting or diffracting objects. The directionality of the received low frequency sound is greatly degraded because of irregular amplitude fluctuations and signal dispersion. The increase in sound scattering and reverberation in forest is also produced by a large amount of stationary heterogeneities produced by micrometeorological instabilities. For animal communication in forest, reverberation imposes additional limitations related to the rate of repetitive frequency modulation (Wiley and Richards 1978).

Data on reverberation in a pine stand (a monoculture on flat ground, *Pinus nigra*, 29 years old, average diameter 16 cm, average tree height 11.20 m, tree density 0.19 trees/m^2, percent canopy cover 79%, canopy area 22 m^2, n = 0.19 trees/m^2, air absorption 0 dB) were provided by Huismann and Attenborough (1991). Reverberation was measured by switching the source on and off. Pink noise was used as the signal. The real-time analyzer measured a large number of one-third octave band spectra in its transient mode (Fig. 4.45). The decay curves measured at 1,000 Hz and 4,000 Hz at distance of 10 m and 100 m from the source showed that the curves coincide shortly after the moment that the source-off event reaches the 100 m receiver (after 320 ms). At the moment

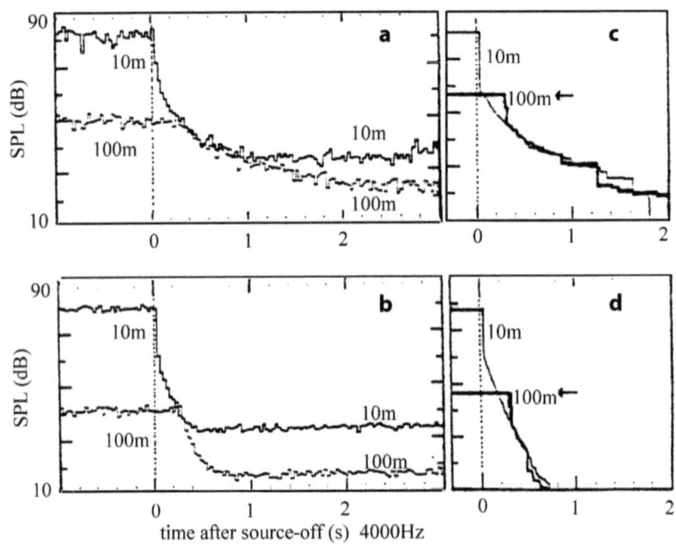

Fig. 4.45. Reverberation SPL at 4,000 Hz as a function of time after the source-off. Measured and modelled noise-off decay for 1 kHz and 4 kHz (Huisman and Attenborough 1991). Reprinted with permission from the Acoustical Society of America, copyright 2005. **a** Simultaneously measured source-off response at 10 m (*upper lines*) and at 100 m (*lower lines*), at 1 kHz. **b** Simultaneously measured source-off response at 10 m (*upper lines*) and at 100 m (*lower lines*), at 4 kHz. **c** Modelled source-off response that fit to experiments in **a**, air absorption 0 dB/s, effective scattering diameter 0.04 m. **d** Modelled source-off response that fit to experiments in **b**, air absorption 10 dB/s, effective scattering diameter 0.16 m

at which the switch-off event reached the receivers, a steep fall in the levels was observed, corresponding probably to the contribution of the reverberant sound field to the total sound field. The high-effective tree absorption at 4,000 Hz means that the nonlinearity of the decay curve almost disappeared (Fig. 4.45b).

Reverberation was measured both for its own interest and in order to obtain data on the scattering effect of vegetation, independent of ground effect and meteorology, using the model proposed by Kuttruff (1967). In this theoretical model, the trees are perfect cylinders and the forest is represented as an isothermal windless air volume without ground, containing a random array of infinitely long parallel cylinders which scatter sound particles from a point source.

Theoretically, the energy in the reverberant field in the pulse response is related to the direct field attenuation, to the total scattering attenuation and to the energy in the field, as shown by (4.16a), (4.16b).

$$A_s = 10\log\left[\frac{(E_d + E_r)}{E_f}\right](dB) \tag{4.16a}$$

$$A_d = 10\log\frac{E_d}{E_f}(dB) \tag{4.16b}$$

Table 4.13. Insertion loss (dB) for barriers with different edge profiles (Ishizuka and Fujiwara 2004). Reprinted with permission from Elsevier, copyright 2005

Barrier and edge profile	Characteristics	Insertion loss	
		Mean value (dB) (IL)	Relative value (dB) (ΔIL)
Plain, rigid surface	3 m reference	15.2	0.0
	6 m	20.1	4.9
	10 m	23.4	8.2
Rectangular edge	Rigid surface	16.2	1.0
	Absorbing	19.7	4.5
	Soft	23.1	7.9
T-shaped edge	Rigid surface	17.1	1.9
	Absorbing	20.5	5.3
	Soft	23.6	8.4
Cylindrical	Rigid surface	14.7	−0.5
	Absorbing	19.2	7.6
	Soft	22.8	7.6
Double cylindrical	Rigid surface	17.9	2.7
	Absorbing	20.4	5.4
	Soft	23.3	8.1

where: A_s is the total scattering attenuation, A_d is the direct field attenuation, E_d is the energy in the direct field, E_r is the energy in the reverberant field and E_f is the energy in the free field.

As can be seen from Fig. 4.45, the measured nonlinear decay of levels with time seems to be consistent with the model proposed by Kuttruff. For quantitative calculations, the values of R_e (effective refraction factor) and D_e (effective

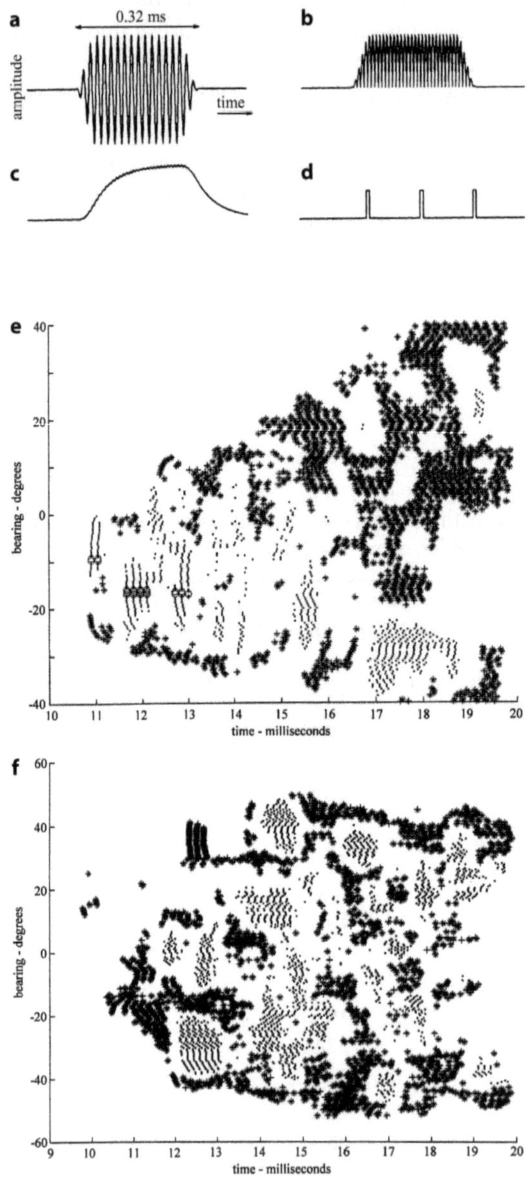

Fig. 4.46. Pseudo-action potentials produced by echo signal processing used for the identification *of Rhododendron maximum* and *Taxus media* (Kuk 2001). Reprinted with permission from the Acoustical Society of America, copyright 2005. **a** Simulated echo waveform stimulus, **b** half-wave rectified waveform–transmitter substance release, **c** output signal at lossy integrator–transmitter, **d** sequence of three pseudo-action potentials. **e** *Rhododendron maximum* located at 1.95 m range with mobile sonar and corresponding to the outputs of classifier neurons, deduced from a pseudo-action potential field produced by a scan plane tilted 35° downward from horizontal. **f** *Taxus media* located at 1.8 m range with mobile sonar and corresponding to the outputs of classifier neurons, deduced from a pseudo-action potential field produced by a scan plane tilted 35° downward from horizontal

trunk diameter) have to be obtained by fitting predictions to the measurements. Table 4.13 shows a good agreement between the theoretical and experimental data only for a limited number of data. The decrease in D_e with increasing wavelength is well comprehensible, but $R_e = 0.1$ is low.

From all these data, it seems clear that there are more phenomena than just scattering that influence sound transmission (ground effect, etc.). An important statement of the modelling proposed by Huisman and Attenborough (1991) is that "the important contribution of the direct field implies that the total attenuation due to scattering A_s can be approximated by the direct field attenuation A_d".

The information content of echoes from in situ plants and trees can be used for their identification (Kuc 2001), by transforming echoes into pseudo-action potentials for classifying plants, using conventional mobile sonar, with a narrow bandwidth of about 3 kHz, operating in metrology and mobile robot applications.

Specular plants, such as the rhododendron (*Rhododendron maximum*), have flat leaves that are large, compared to the wavelength, and act as isolated specular reflectors. The sonar response is characterized by large amplitude and coherence over successive echoes. Diffuse plants with needles, such as yew (*Taxus media*), act as diffuse scatterers and produce many small echo components, which superimpose randomly at the detector. This phenomenon is produced because needles are small compared with the wavelength. The information content of point process extracted from in situ measurements, through their characteristic pattern, also identified trees such as sycamore (*Platanus occidentalis*) of 17 cm diameter and maple (*Acer platanoides*) of 80 cm diameter. Figure 4.46a–d shows the echo processing to produce pseudo-action potentials. By rotating the transducers while emitting probing pulses and processing the echoes, the sonar forms a sector scan of the environment. From one emission, the signals have a temporal response; and from the sector scan, the signals have a spatial dimension. The echoes used as navigation landmarks were classified using delay and coincidence detection. The sequence for *Rhododendron maximum* is shown in Fig. 4.46e and those for *Taxus media* in Fig. 4.46f.

4.3.4 Atmospheric Conditions

Outdoor acoustic measurements are affected by the atmospheric conditions, relative humidity, temperature or wind gradients (Beranek 1971). The structure of the atmosphere varies with climate, weather, local conditions, diurnal cycles and insolation. All these factors influence sound transmission in a random way (Brown 1987). The sound speed profile under different meteorological conditions has been widely studied (Waxler 2004) for various applications including physical meteorology, community noise modelling, bioacoustics and forestry. The meteorological conditions have a screening effect on sound and noise transmission (Barrière and Gabillet 1998). The wind rustling sound emitted when vegetation rustles in the wind is perceived by humans as a comfortable sound. Yamada et al. (1977) noted that the rustling sound is very similar to white noise, having the same octave bands.

Inside the forest, wind speed and temperature gradient are reduced and, because of this, the total acoustical efficiency increases in the case of favorable propagation conditions. Wind penetrating direction has an important impact on wind speed in forest. In the trunk space, the wind speed is greater than in the canopy – variations of 45% and more can be expected (Raynor 1971). Huisman et al. (1988) reported data on temperature variation in a forest stand (Fig. 4.47) and noted that the crown layer is thermodynamically active. Near the ground, temperature effects are neglectable up to at least a distance of 100 m, because of the quiet isotherm air layer below the canopy. A ray tracing calculation could demonstrate that, depending on the shape and profile of the canopy, the sound refracted by the crown layer came down to a distance of a few hundred meters. Ohsato (1972) demonstrated the wind masking effect on a 2 kHz pure tone propagation in forest.

"In the forest, the effect of the fluctuating sound velocity profile in the late morning and the afternoon on sound transmission was clearly audible, especially if pure tones of chords were used as signals. The effect is an apparently random fluctuation of SPL of any single tone but with a variable amount of correlation between the various tones of the chord. These fluctuations can be large, for ex. at a distance of 100 m on a bright day; we monitored a change in level of a 4 kHz pure tone of 11 dB within a period of 50 s. With one-third octave band noise, these effects were less pronounced, which indicates that

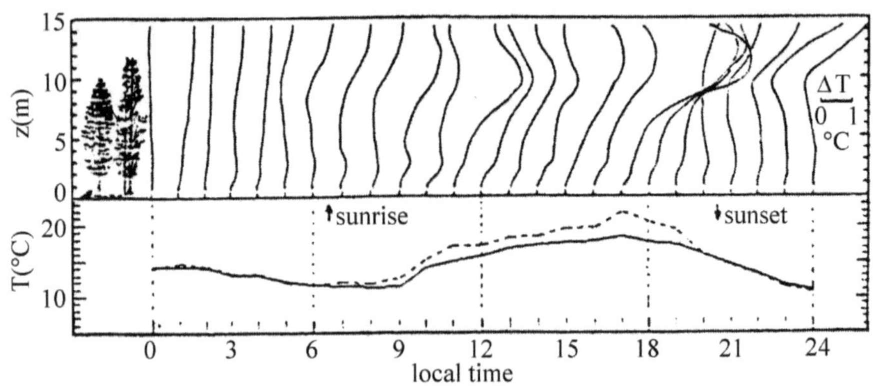

Fig. 4.47. Wind and temperature profiles during the day time (Huisman and Attenborough 1991). Reprinted with permission from the Acoustical Society of America, copyright 2005

interference plays a role" (Huisman and Attenborough 1991). This is a very valuable observation in forest, since it motivates subsequent investigations. Using a numerical simulation of wind and sound propagation through an idealized stand of trees, Heimann (2003) demonstrated that direct attenuation by trunks (produced by multiple reflections and scattering) is much larger than indirect attenuation, which is due to a reduction of the vertical wind gradient in the stand.

A comparison between winter and summer measurements in different forest sites allowed an assessment of the effect of season and foliage on sound attenuation (Price et al. 1988). The attenuation is significantly less in winter, because of the absence of leaves. A peak of attenuation observed at 200 Hz is attributed to the ground effect; and the gradually increasing attenuation for frequencies higher than 1 kHz is attributed to scattering both by trunks and by foliage (see previous Fig. 4.19).

Relative humidity affects the properties of all porous materials and consequently the impedance of the ground and the absorption and scattering by trunks, foliage or bark. The effect of relative humidity on attenuation in a pine plantation is shown in Fig. 4.48 over a period of 2 months (during which time no rain was recorded). It appears that the relative humidity of the atmosphere has a very important effect on the measured attenuation rates, at all frequencies. The maximum attenuation rate occurs at about 75% relative humidity.

Atmospheric absorption as a function of relative humidity is shown in Fig. 4.49. The variations in the absorption coefficient expressed in dB per 100 m at normal atmospheric pressure and 20 °C are due to molecular absorption

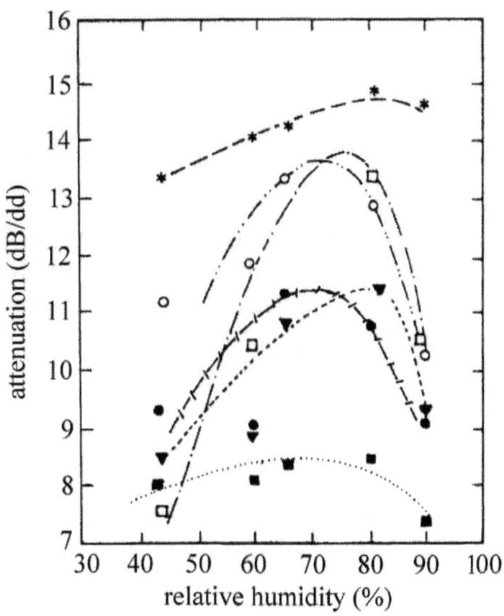

Fig. 4.48. Effect of air relative humidity on attenuation in a pine plantation for different frequencies (Fricke 1984). Reprinted with permission from Elsevier, copyright 2005. *Circles* 31.5 Hz, *black squares* 125 Hz, *black circles* 500 Hz, *black triangles* 1 kHz, *white squares* 2 kHz, *stars* 4 kHz

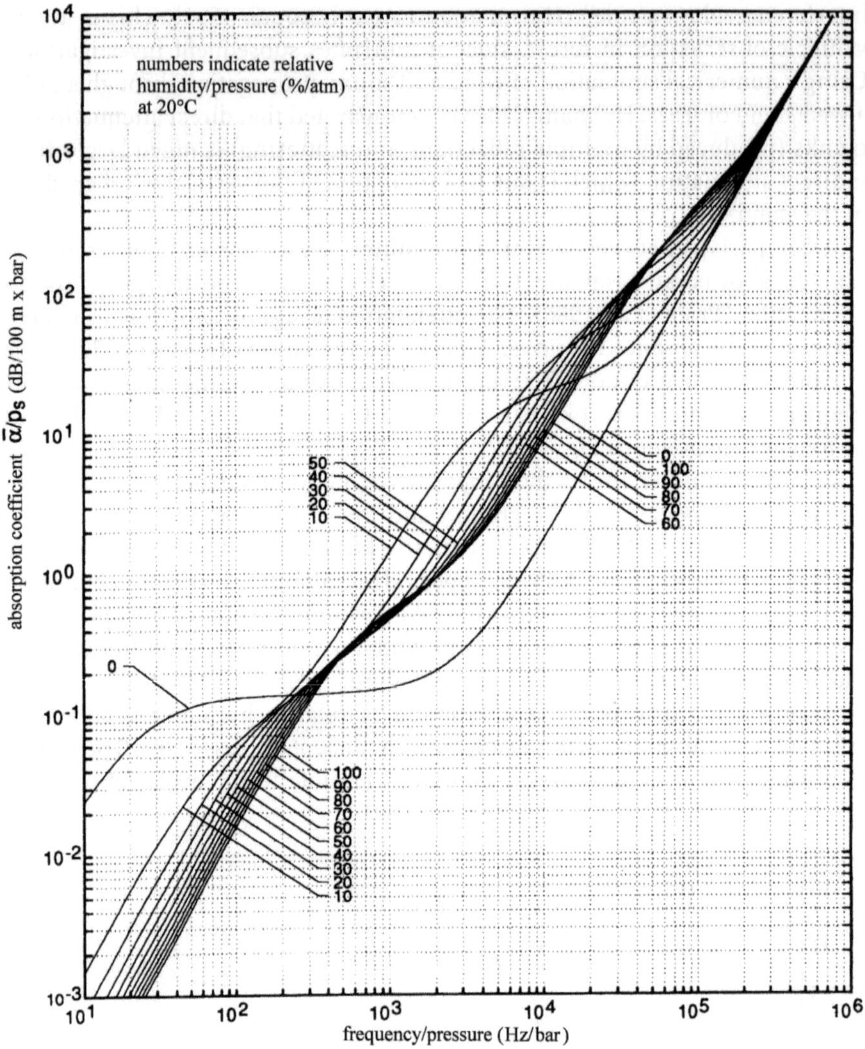

Fig. 4.49. Air relative humidity in air due to molecular absorption at 20 °C. The *ordinate* gives the rate of conversion of sound energy into heat during sound propagation and is expressed in dB/100 m. The *abscissa* is the ratio between the frequency from 10 Hz to 1 MHz and the pressure (Hz/atm; Bass et al. 1995). Reprinted with permission from the Acoustical Society of America, copyright 2005

and depend on pressure. The molecular absorption of oxygen and nitrogen molecules converts a very small fraction of the acoustic wave energy. Above 500 Hz, the predominant mechanism is related to the oxygen–water vapor molecular relaxation. This mechanism induces an attenuation of 2 dB/km. This attenuation can increase with frequency. Below 500 Hz, the nitrogen–

water vapor relaxation is observed, producing less energy absorption then the oxygen–water vapor relaxation (Embelton 1996).

For a better understanding of the complex phenomena related to sound propagation in a forest, modelling and simulations must be used. For this purpose, several acoustical parameters such as attenuation, reverberation and echoes can be used. Acoustically relevant characteristics of a forest are: tree species, trunk diameter, number of trees/unit surface, scattering and absorbing cross-sections, leaf area, mean free path length, visibility and light penetration path. The propagation of sound through an idealized stand of trees was modelled by Heimann (2003) in 3-dimensions, using a numerical finite difference time–domain fluid mode, with the discussion concentrated on attenuation. Systematic simulations performed by varying the number of trees/unit surface and the trunk diameter showed that the simulated attenuation by the trunks agrees with scattering theory. Direct attenuation by the trunks through multiple reflections and scattering is much larger than indirect attenuation due to the reduction of the vertical wind gradient in the stand and the corresponding reduction in acoustic wave refraction.

4.4 Sound Scattering by Barriers

4.4.1 Psychological Effect

The variability of individual responses to a given noise exposure is remarkably high. There appears to be a widespread popular belief that belts of trees can cause an important reduction in traffic noise (Watts et al. 1999). Survey respondents residing close to roads express the wish that bands of trees should be used to screen traffic noise along departmental roads. All over the world, owners plant hedges or trees in their gardens to screen traffic noise. Huddart (1990) measured a reduction of 6 dB(A) produced by a densely planted belt of trees 30 m thick, compared with grassland of the same thickness. Perfater (1979) noted that, when existing vegetation along a road was replaced by a solid barrier, the residents clearly felt that the vegetation had given better noise reduction. It seems that, sometimes, the effect related to the attractive visual appearance is predominant over the effective acoustical benefit.

It was proved that, when the ambient noise level was held constant (a single tone at 500 Hz which was varied between 50 dB and 80 dB and replayed through headphones), the loudness increased as the percentage of vegetation increased (Mulligan et al. 1987). This psychological effect can be explained by the fact that, when the source is visually screened, a listener expects its loudness to be significantly diminished.

Aylor and Marks (1976), using a white noise source limited by filters to a one-third octave band centered at 1,000 Hz, with a sound pressure level variable between 40 dB and 100 dB, found that a thick conifer hedge or louvered barrier with gaps (which allowed the direct transmission of sound but which completely obscured the source) produced similar sensitivities. When the source was visible either through an open-slat fence or where there was no screen at all, sensitivities were significantly lower. The difference at 65 dB sound pressure level for the open fence and conifer hedge was 7.5 dB; and for conifer hedge and no screen, the difference was 3.5 dB. It was noted that the effect was purely visual and was not due to the differences in the frequency spectra of noise transmitted through or diffracted over the different barriers. These effects are comparable with the differences reported by Watts et al. (1999) between a willow barrier, metal barriers and without a screen.

Aylor and Marks (1976) blindfolded a sample of listeners. They were not able to observe differences in the ratings produced for the different listening conditions.

Watts et al. (1999) demonstrated that the presence of vegetation between the source and receiver had little effect on the sound spectrum, but the noise barriers reduced the levels at higher frequencies. The A-weighted level was not an adequate measure of noise exposure, since it might have failed to reflect these differences in the spectral balance. A more sophisticated parameter which takes account of such changes is the loudness level.

If the rate of the attractiveness of the barriers is scaled from 0 to 9, where 0 corresponds to “very unattractive” and 9 corresponds to “very attractive”, a metal barrier 15 m long and 3 m high is rated at 2 and a woven willow vegetative noise barrier with earth fill, of the samesize, is rated at 7. This was not apparent from the noise sensitivity results. It was supposed that the willow barrier effect was perceived through the well known “halo” effect and it was concluded that, in this particular case, the effect of vegetation on noise sensitivity was associated with the degree of visual screening of the source. Noise reduction is small unless the vegetation belt is wide.

4.4.2 Solid Barriers Without Vegetation

Solid barriers reduce noise in two ways, by reflecting or absorbing noise. Reflecting barriers are build of any dense material. Absorbing barriers have a perforated skin and a chamber behind, into which the sound waves are dispersed. The simplest environmental barrier is the earth mound. The architectural morphology of barriers must be integrated into the local landscape. The good design of environmental barriers must take into consideration solid materials as well as the plants which help to integrate barriers into their surroundings, reducing their apparent scale and softening their appearance by providing robust features.

The most common noise barrier is composed of a single vertical screen which obstructs the propagation of sound from source to receiver. Adding side panels to this profile or using two or more screens could be considerably more efficient in terms of noise reduction.

The field performance of noise barriers has been evaluated for more than 30 years (Raynor 1971; Scholes et al. 1971; DeJong and Stusnick 1976; Reethof and Heisler 1976; Watts et al. 1994; Jimenez-Altamirano 1997; van Renterghem and Botteldooren 2002, 2003).

Classic barrier modelling in a homogeneous atmosphere is shown in Fig. 4.50, in which the diffraction sound path around corners over wide barriers is traced. The barrier insertion loss prediction provides an accurate means

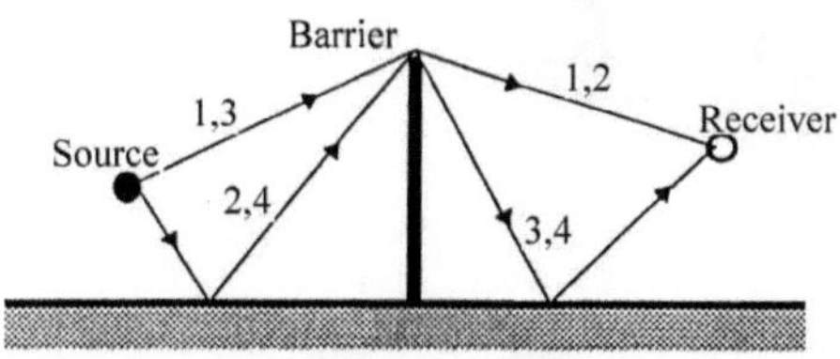

Fig. 4.50. Diffraction path induced by the simplest noise barrier (Muradali and Fyfe 1999). Reprinted with permission from Elsevier, copyright 2005

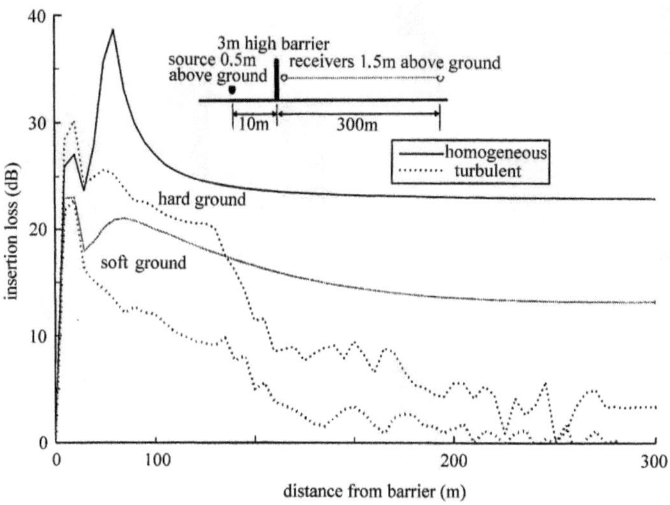

Fig. 4.51. Theoretical insertion loss produced by a barrier of 3 m height, at 500 Hz for asoft and hard ground (Muradali and Fyfe 1999). Reprinted with permission from Elsevier, copyright 2005

to consider reflection, diffraction and phase interference in the sound field around the barriers.

Figure 4.51 shows the insertion loss as a function of distance for a 3 m barrier height over rigid and soft ground in a homogeneous and turbulent atmosphere. It can be seen that the insertion loss is lower for the soft ground case. The sound absorption expressed by the insertion loss greatly deteriorates after about 75 m which means that, after this distance, the rays that pass over the barriers negate most of the barrier shielding.

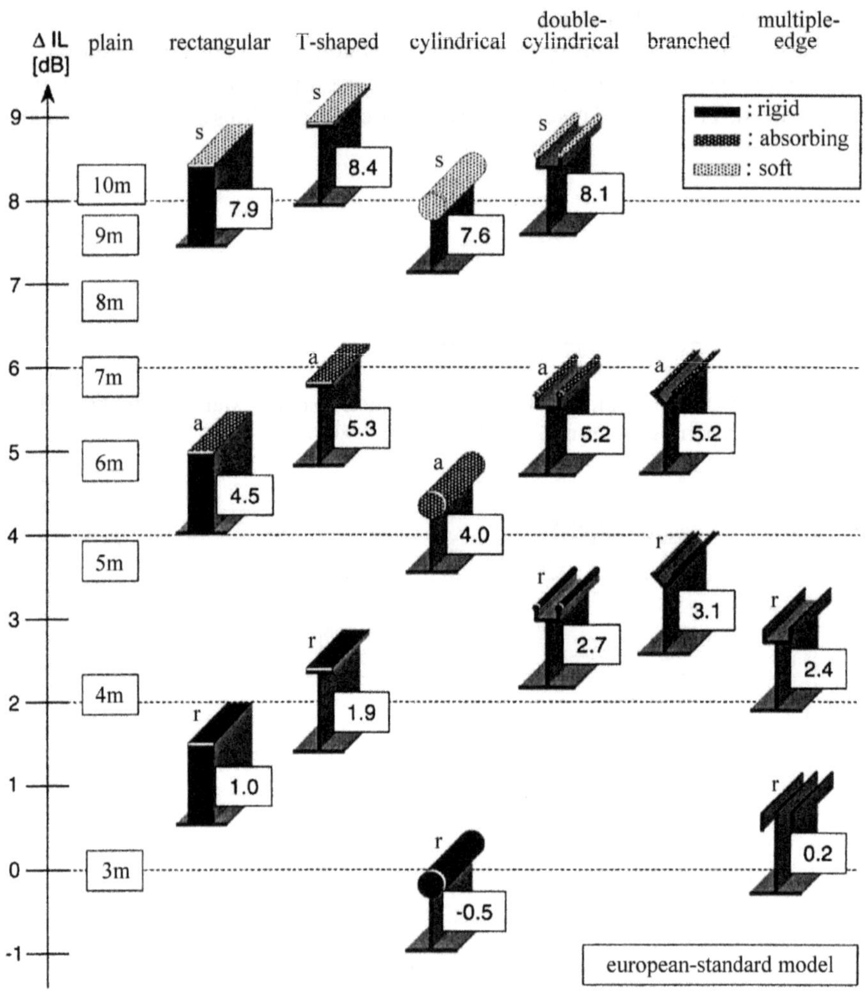

Fig. 4.52. Comparison of the performance of barriers (ΔIL in dB) with various shapes (plain, rectangular, cylindrical, etc.) for the European standard model (Ishizuka and Fujiwara 2004). Reprinted with permission from Elsevier, copyright 2005

Several technical solutions exist for improving the acoustical performance of plain barriers, such as modifying the shape by introducing two or more diffracting shaped edges or by suppressing the sound pressure at the edge by installing a soft absorbent material and decreasing the diffracted field behind. The shaped edges of noise barriers are very complex and can be T-shaped, cylindrical, double-cylindrical, branched, multiple-edged, or with side panels (Ishizuka and Fujiwara 2004). Figure 4.52 shows a comparison between the performance of barriers with various shapes for the European standard noise source model EN 1793-3. The reference is a 3-m high plain barrier which is compared with plain barriers of different heights varying from 0.2 m to 8.0 m. All tested barriers had a maximum width of 1 m. The mean insertion loss (*IL*) was measured for a broadband noise spectrum under the "shadow" region of the barrier. The losses were measured at six positions and averaged. The efficiency of barrier shape for three surface conditions (rigid, absorbing, soft) was expressed by ΔIL, which is the difference between the tested barrier and the reference plain barrier of 3 m height. Data from Table 4.13 show the important efficiency of rectangular, cylindrical and double-cylindrical barriers, for which ΔIL is between 7.6 dB and 8.4 dB.

The acoustical performance of noise barriers can be determined using different techniques, such as diffraction techniques based on geometrical ray theory, boundary element method, finite element method and finite wave envelop method.

The reader interested in these theoretical methods can refer to Crombie et al. (1995), Duhamel et al. (1998), Muradali and Fyfe (1999).

4.4.3 Solid Barriers with Vegetation

Along many road corridors, the introduction of environmental noise barriers which fit with the local environment was effective with the introduction of noise legislation. This integration into the local landscape or townscape can be achieved by planting with different species and in this way can soften the appearance of the barrier by breaking the scale of observation. Figure 4.53 shows the increasing attenuation induced by the combined action of trees and barriers.

The simplest effective environmental noise barrier is the earth mound. A comparison between the land-take for a 4 m high earth mound and a 4 m high bio-barrier is given in Fig. 4.54. The bio-barrier needs a space only 2.5 m wide, while the earth mound is positioned on a space 14 m wide. Planting with vertically oriented species like for example *Parthenocissus quinquifolia* does not require a large amount of space (30 cm) and improves the appearance of the barrier. For woody plants, 1 m is required (Kotzen 2004).

Bio-barriers which incorporate planting within their structure are shown in Fig. 4.55. Following their shape, these bio-barriers are classified as:

1. A-frame barrier, having an internal tie, with the plants affixed by rubber straps;
2. Vertical barriers on timber supports and rubber straps;
3. Box wall, on a steel support and appropriate foundations, a welded steel grid, with a horizontal strut and an irrigation pipe;
4. willow-weave wall on a geogrid and ropes;
5. stack with steel and concrete struts and the planting medium inside the structure.

The materials and technical solutions used for different barriers are shown in Fig. 4.56. Planting behind and back to the working face of the barriers is always important for perfect integration of the barriers into the landscape. The skilful design of planted barriers is an outstanding example of acoustic ecology, encouraging a deeper appreciation for noise and its role in our lives.

A combined effect on excess attenuation of a barrier (2 m height) and belt of trees is shown in Fig. 4.57, for traffic noise on an asphalted road. A reduction of about 3 dB(A) can be observed in a zone between 100 m and 450 m distance from the source and 1,030 m height, probably due to the canopy. The iso-excess

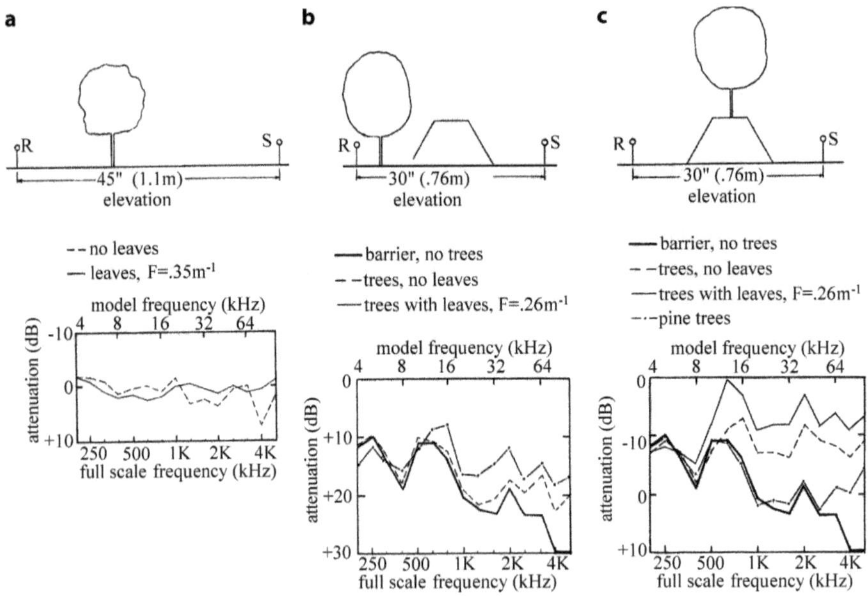

Fig. 4.53. Excess attenuation barriers and trees. **a** Tree on a flat plain case, **b** barriers with trees along the side, **c** trees on the top of the barrier (Lyon et al. 1977)

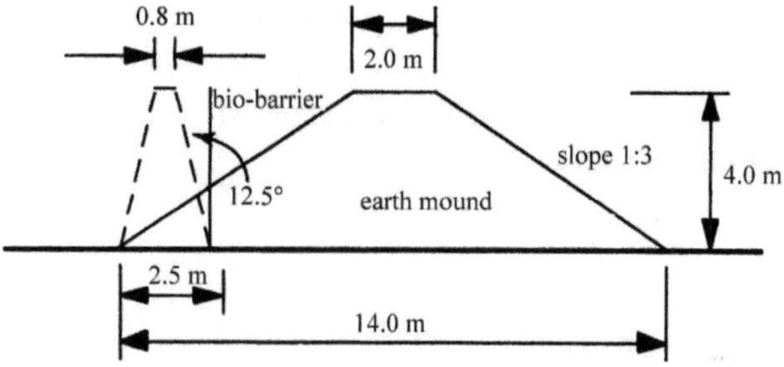

Fig. 4.54. Comparison between bio-barrier and earth mound of the same height (4 m) and of different wide (2.5 m for the bio-barrier and 14 m for the earth mound; Kotzen 2004)

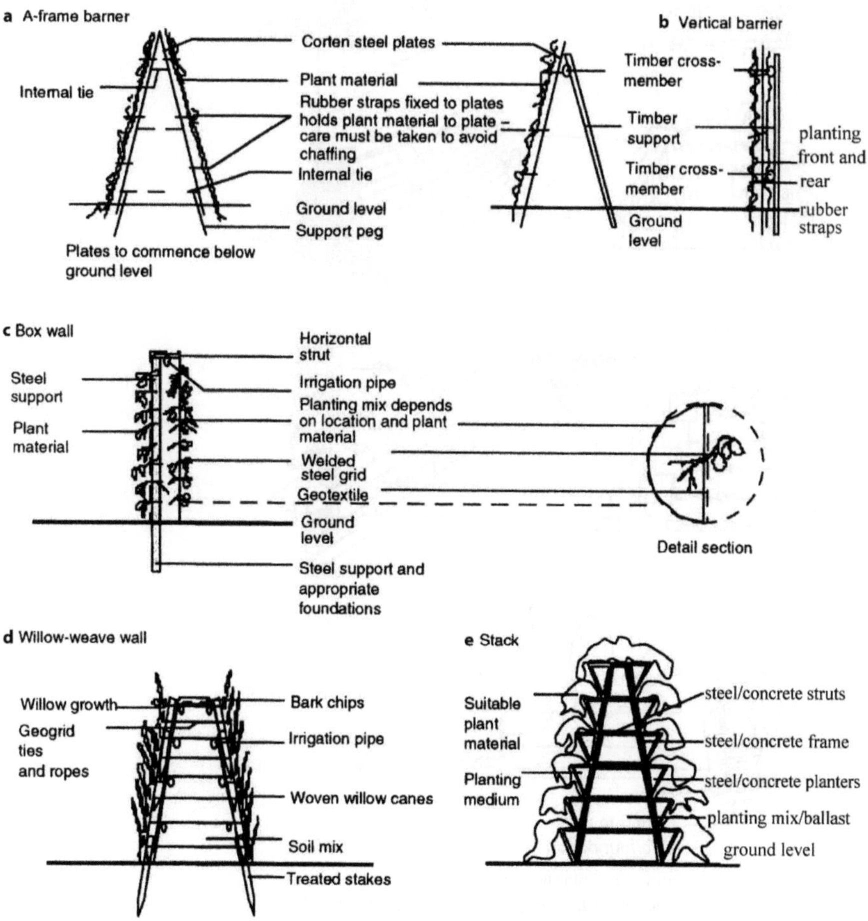

Fig. 4.55. Bio-barriers incorporating planting (Kotzen 2004). Different types of barriers: **a** A-frame barrier, **b** vertical barrier, **c** box wall, **d** willow-weave wall, **e** stack. Materials used are timber, concrete, earth

attenuation of 2 dB(A) between 100 m and 600 m is probably produced by the scattering of trunks. Under the canopy, between about 1 m and 6 m height, the iso-excess attenuation is very low (less then 1 dB). The barrier effect (4 dB) can be seen at 300 m distance from the source.

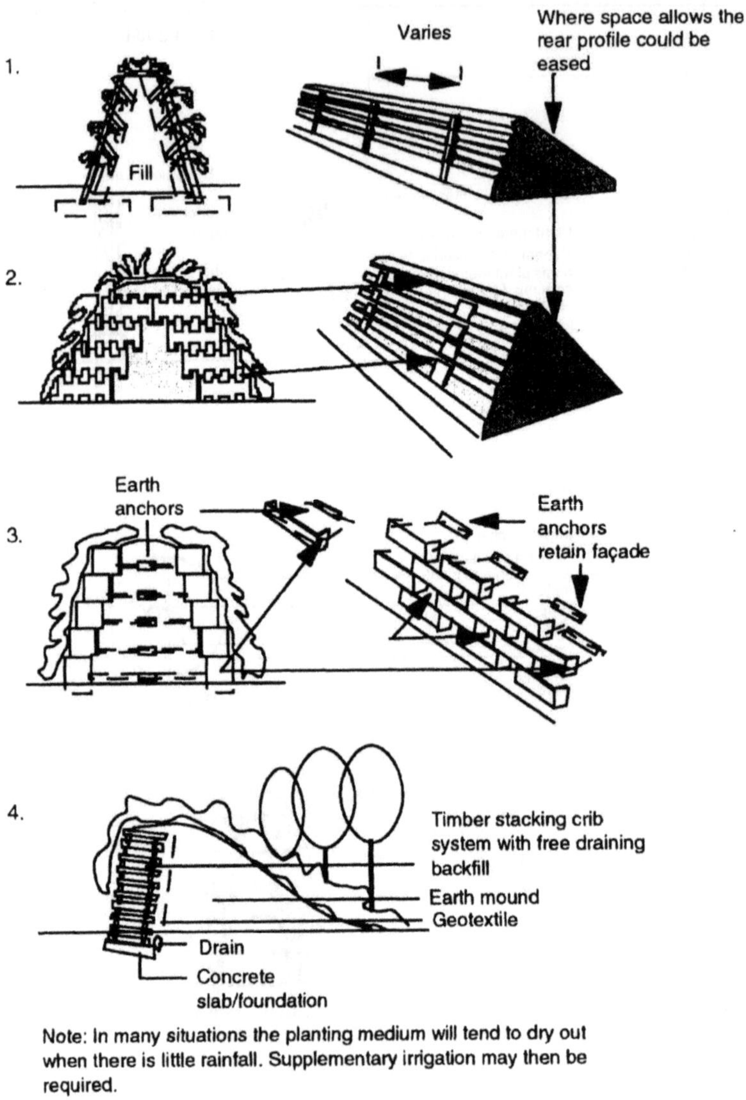

Fig. 4.56. Different materials (timber, concrete, earth) used for bio-barriers (Kotzen 2004)

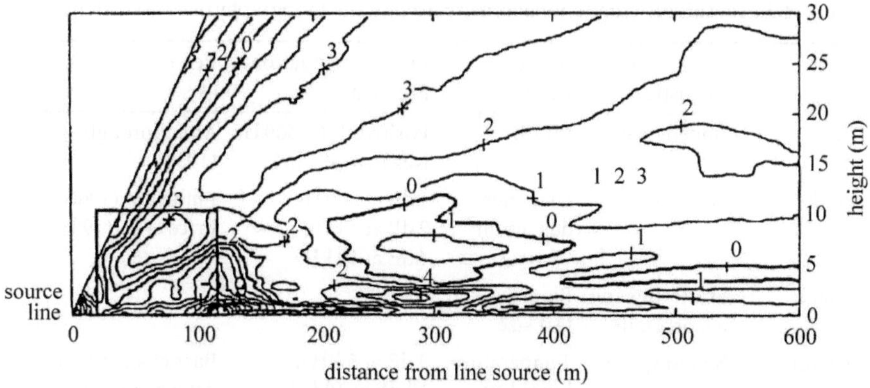

Fig. 4.57. Iso-excess attenuation curves [dB(A)] as a function of the distance from the line source and the height of the trees (Barrière and Gabillet 1998)

4.5 Summary

Plants in general can attenuate sound by reflecting and absorbing energy. Sound transmission by plant material, namely trees, tree belts and forest stands, is expressed as excess attenuation, i.e. the measured sound pressure level corrected for air absorption, minus the free-field level. The free-field level is the level that would exist if there were no obstacle and no sound velocity gradient. The attenuation properties of vegetation are questionable because of the great number of variables involved. Certain types of vegetation are better for attenuating sound than others. A valid comparison between the performances of different types of vegetation cannot be made without an exact description of the methodological factors characterizing the noise source and the receiver, such as height, placement of source (whether inside forest or outside and, if outside, how far outside) spectrum of source and its duration (steady or transient), size and density of trees and atmospheric conditions during experiments (temperature and wind gradients, relative humidity; Table 4.14). The ground is a significant absorber of sound in forest. The soft forest floor has a pronounced influence on low-frequency sound propagation. An excess attenuation of 6 dB per doubling of distance is shown to be possible.

Acoustic characteristics of the constitutive elements of the tree, such as trunk, bark and canopy, and of the forest floor should provide needed insight into the acoustic absorption mechanisms. An acoustic wave of 1 kHz frequency propagates with a wavelength comparable with the trunk diameter. The trunk, branches and foliage partially scatter the incident acoustic energy. Scattering effectiveness is consistent with the geometry of the scatterers, the bigger the scatterers, the lower the frequency at which the scattering phenomenon be-

Table 4.14. Effects of soil, trunks, foliage, meteorology and topography on excess attenuation

Effect	Characteristics		Excess attenuation	References
	Acoustic	Other	Frequency	
Soil	Impedance	Width	Positive 100–500 Hz Negative at 1 kHz	Attenborough (1985)
Trunks	Scattering	Diameter, density/m^2	1 dB at 300 Hz 2 dB at 1 kHz 4 dB at 10 kHz	Embelton (1966), Price et al. (1988)
Foliage	Scattering and absorption	Biomass, leaf size		
Meteorology	Sound speed	Temperature, humidity	3 dB at 5 kHz, 10 dB at 10 kHz, at 100 m, 20 °C	Bass et al. (1995), Attenborough et al. (1995)

comes effective. The audible reverberation in a forest can be explained by the interference between direct and ground-reflected sounds and the scattering effects induced by trees, branches and meteorological conditions. Modelling and simulation can be used for a better understanding of the complex phenomena related to sound propagation in a forest. For acoustics, the relevant characteristics of a forest are tree species, trunk diameter, number of trees/unit area, scattering and absorbing cross-section, leaf area, visibility and light penetration path.

Sound scattering by belt of vegetation and barriers is largely used for reducing traffic noise by reflection and absorption. The skilful design of barriers is an outstanding example of acoustic ecology, encouraging a deeper appreciation for noise and its role in our lives.

5 Traffic Noise Abatement

As is generally accepted, the negative perception of noise is defined as annoyance and the positive perception of noise is defined as comfort. Today there is a need to handle noise at source and to rate it correctly, exploring the physical and psycho-acoustic aspects and also the impact of visual aesthetics on the soundscape (Canévet 1996; Fyhri and Klaeboe 1999).

Noise has an important environmental impact with short- and long-range effects on human communities, on nature and on wildlife (Carlson et al. 1977; Marquis-Favre et al. 2005). Ambient noise is "the all-encompassing noise associated with any given environment, and is usually a composite of sounds from many sources – near and far" (Harris 1998). A description of the noise environment in the audible range, outdoors, is provided by the A-frequency weighted day and night average sound level. The noise description should include a map (Piccolo et al. 2005), drawn in increments of 5 dB, representing the contours of constant values of the yearly day–night average sound level or day–night sound exposure, for the existing conditions. Different methods, which relate the human response to noise, have been used to ascribe a numerical degree of impact on the population (loudness, annoyance, speech interference, hearing loss, etc.). As noted by Bishop and Schomer (1998; Table 5.1), the range of variation in outdoor day–night average sound levels in urban communities is very big. Figure 5.1 shows the A-weighted sound level in early afternoon and

Table 5.1. Variation of the outdoor day–night average sound levels in urban communities in the United States (Bishop and Schomer 1998). Reprinted with permission of the Acoustical Society of America, copyright 2005

Sound level (dB)	Number of people (millions)					
	Traffic only	Traffic and aircraft	Traffic and construction	Traffic and rail	Traffic and industrial	Total
> 80	0.1	0.1	–	–	–	0.2
> 75	0.9	0.5	–	0.1	–	1.5
> 70	4.5	2.2	0.2	1.0	0.2	8.1
> 65	15.2	7.6	0.8	3.0	1.2	27.8
> 60	36.6	16.1	2.8	4.4	3.7	63.6
> 58	49.2	24.3	6.0	6.0	6.9	92.4

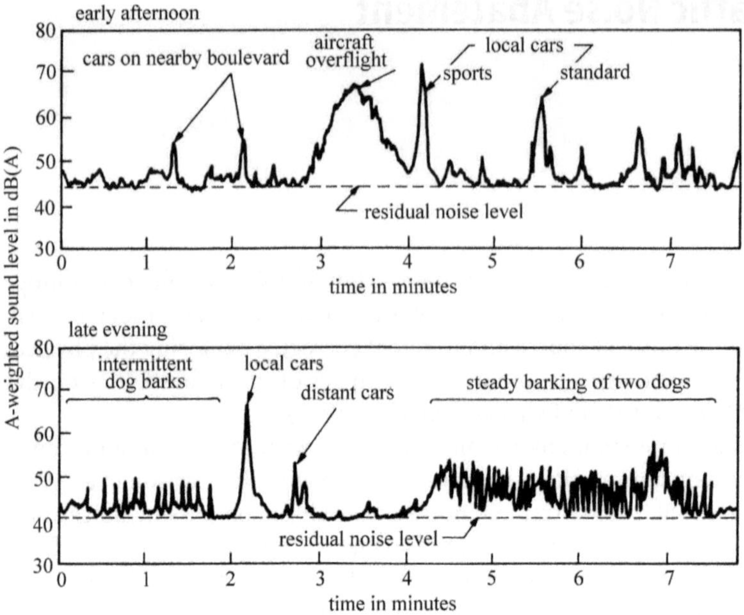

Fig. 5.1. Early afternoon and late evening measurements of A-weighted sound levels of outdoor noise versus time in an suburban area, using a microphone located at 6.1 m from the street curb (Bishop and Schomer 1998). Reprinted with permission of the Acoustical Society of America, copyright 2005

late evening in a residential area. Distinct noise events can be observed in this figure, such as for example local cars, aircraft overflight, or intermittent dog barks.

The Green Paper on future noise policy (European Commission 1996) noted that "the environmental noise, caused by traffic, industrial and recreational activities, is one of the main local environmental problems in Europe and the source of an increasing number of complaints for the public". Over the years, since the second half of the 20th century, a considerable amount of research and development activity was carried out into the noise problem. In more than 50 years, noise at source has been reduced by a considerable amount. Numerous standards and regulations impose noise control and limitation. At the same time, the traffic volume is increasing continuously, as well as people's expectation for quietness. Noise annoyance is fundamentally a matter of public perception. As an example, Table 5.2 gives an assessment of the overall impact of urban traffic noise in excess of 55 dB, in the United States. A reduction of 5 dB in the day–night level represents an improvement of 68%, while a reduction of 10 dB represents an improvement of 89%.

Noise acceptability criteria are difficult to establish, because so much depends on the circumstances of each individual case, which also supposes a sub-

Table 5.2. Overall impact of urban traffic noise on population in the United States (von Gierke et al. 1998). Reprinted with permission of the Acoustical Society of America, copyright 2005

Estimate of current conditions		Annoyance-weighted population			Population-weighted day–night sound exposure (1,000 Pa s^2)		
Day–night contour levels	Population within each contour	Current	For day–night level reduction of:		Current	For day–night level reduction of:	
dB	Millions	0 dB	5 dB	10 dB	0 dB	5 dB	10 dB
50–55	37.0	3.3	1.8	0.7	228	72	72
55–60	34.7	6.3	3.1	1.7	674	213	67
60–65	17.4	5.6	3.1	1.6	1,070	338	107
65–70	5.6	3.0	1.8	1.0	1,088	344	109
70–75	1.2	1.0	0.7	0.4	738	233	74
75–80	0.1	0.1	0.1	0.1	194	62	19
Total	96.0	19.3	10.6	5.5	3,992	1,262	448
Noise impact index		0.2	0.1	0.06	–	–	–
Reduction	–	–	–	–	–	68%	89%

jective approach. In most practical situations, the costs of noise control versus the environmental consequences must be taken into consideration. Most residential environments are affected by more than one noise source. Decreasing the level of a steady noise of 1 dB or less is detectable only in laboratory conditions. Noise annoyance generally increases with noise level. Variations of 5 dB are realistic under real-life conditions, for which the noise sources (aircraft, railway or road traffic) are not steady. Different technical methods of assessment are available; and it is important that these should be carefully selected to mach the precise objectives of the assessment in each particular case. Always, it is useful to support the best compromise solution that can be achieved.

The environmental noise impact depends on the total energy received at the observation point, the rate of occurrence of noise events and the magnitude of any noisier single event. The statistical description of community noise (Kinsler et al. 1982) is based on the A-weighted measurement of different parameters, as noted in Chap. 4.

The highest community noise rarely exceeds 80 L_{eq} or 120 $L_{A_{max}}$ (the maximum time-weighted level). Typical levels of community noise caused by traffic or noisy neighbors are in the range of 45–75 $L_{A_{eq},16h}$ (for 16 h time-long term linear averaging).

The reduction of noise level requires the following three main steps: the evaluation of the noise environment under existing conditions, the determination of the acceptable noise level and the determination of the difference between the two previous steps.

In urban communities, the sources of noise are numerous and may include noise produced by highways, rail and aircraft transportation. The techniques for noise control are related to control at the source, at the receiver and at the transmission path. The source, the receiver and the transmission path are continuously interrelated. The output of the noise source, which is never constant, depends on the environment in which the source is located. At the same time, the response of the receiver depends upon the characteristics of the path and source.

Noise control at the source can be operated by reducing the amplitude of the vibrations, by reducing the motion of the components into vibration, or by using damping materials. Noise reduction can also be obtained by controlling and reducing the transmission path:

- by reducing the energy transmitted to the receiver (for example: by increasing the distance between the source and the receiver, by altering the relative orientation of the source and receiver, by taking advantage of the natural topography and wind, etc.);
- by introducing barriers of a large size compared with the wavelength of the noise source (for example: by reflection, the noise field of jet aircraft engines can be oriented toward the sky);
- by enclosures around the noise source and receiver.

Noise abatement measures and noise control in communities and industries has become in large part a matter of law. The noise is a parameter of the quality of the acoustical environment which has to be included into the planning process and to be analyzed in an environmental impact statement.

The main outdoor sources of noise in modern life, which significantly affect the quality of human environment, are generated by traffic on the streets or highways, by rail transportation and by aircraft. The synthesis of social surveys on noise annoyance published by Schultz (1978) demonstrated that correlations exist between the degree of exposure to the noise and the intensity of annoyance felt by the subjects.

A combination of plant materials (trees, belt of trees, shrubs and a soft surface of grass or other plant materials throughout the area) and specific topographic situations may provide some immediate improvement in noise abatement, with the likelihood of better conditions as the plants mature.

5.1 Road Traffic Noise

Models of different complexity (Attenborough 1982; Steele 2001) have been used for traffic noise prediction since 1950. Earlier models (which are obsolete today) were designed to predict a single vehicle sound pressure level (L_{P}) at the roadside, with the assumption that the vehicle had a constant speed. Later models were developed for the equivalent continuous level (L_{eq}) for traffic over a chosen period, under interrupted and varying flow conditions, with linear, A-weighted levels and one-third octave band spectra. The studied sources were single points, short line sources, double point sources, or multiple point sources with different spectra. As suggested by Steele (2001), the ideal model proposed a source composed by a multi-stream (branched and interlocked streams of vehicles), with a source/receiver weighting, with octave/dB(A) determining L_{eq}, L_N, L_{min}, L_{max}.

The tendency to unify noise calculation algorithms in Europe was realized with the standards ISO 9613-2 (1996) and EN 1793-3 (1997). A procedure to find automatically all relevant sound paths in an arbitrary two-dimensional terrain for road traffic noise was developed by Heutschi (2004). A-weighted excess attenuation of a highway traffic line source with ground type was discussed by Attenborough (1982). Meteorological factors and ground conditions influence sound propagation over open ground and it is difficult to differentiate ground effects from other outdoor measurements. The forest floor is a soft floor for which the excess attenuation increases with the horizontal distance from source line to receiver.

A typical situation for the description of road traffic noise and the possible attenuation by a belt of trees and barriers is synthesized in Fig. 5.2. The first option uses only distance as a means of noise attenuation. The second option uses tunnels and a belt of vegetation. The third option uses the natural site topography, creating a corridor for automobile and truck traffic. The belt of trees is between the corridor and the buildings. The fourth option uses earth mounds, on which trees are planted. The fifth option uses different types of barriers. The following three options are combined solutions for traffic noise reduction. The attractive visual appearance of a belt of trees can affect the perception of traffic noise (Watts et al. 1999).

Heavy trucks are the most important noise generators on highways. Table 5.3 gives the octave-band sound pressure level for heavy trucks and automobiles for two speeds (56 km/h and 88 km/h) measured at 1.2 m above the ground and 15 m distance from the source (Bowlby 1998). The heavy trucks travelling at 88 km/h produced an A-weighted sound level of 87.5 dB(A), which is about 20% higher than that produced by automobiles travelling at the same speed.

Figure 5.3 gives data on traffic noise reduction by two belts of trees, (deciduous trees and bushes), one of 19 m width, the other of 25 m width, compared with a corn field. The excess attenuation in one-third octave bands shows a maximum around 500 Hz, which is probably due to the interference between the directly transmitted wave and the reflected wave from the ground and the belt of trees. The increasing attenuation between 2 kHz and 4 kHz can be attributed to the branches and leaves (Kragh 1981). In this experimental configuration, no significant differences were observed between $L_{A_{eq}}$ at 1.5 m above

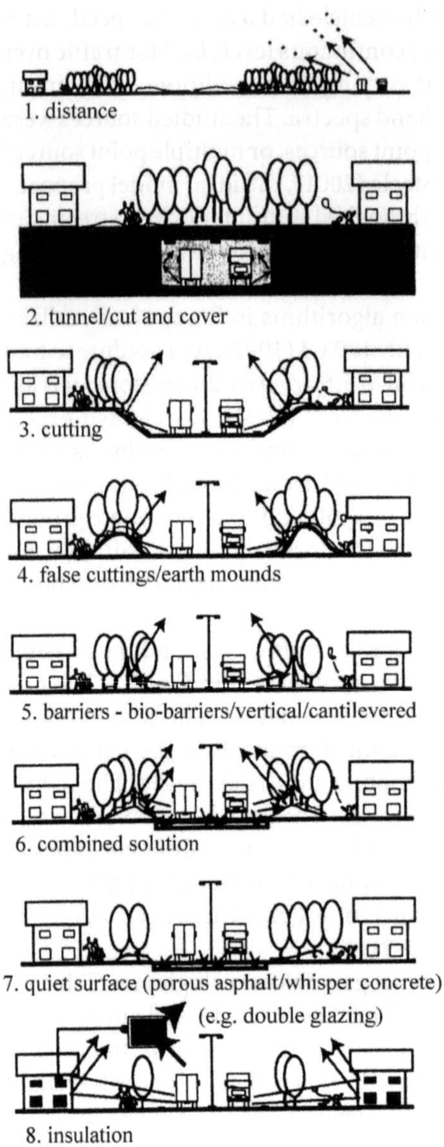

Fig. 5.2. Road traffic attenuation with belt of trees and different barriers (Kotzen 2004)

grass-covered ground and through the belt of trees, probably because only one parameter is not sufficient to express the complexity of the experimental situation.

Figure 5.4 shows the propagation of noise over a forest stand composed of parallel bands of Norway spruce and red pine and having the edge limited by a road and a row of deciduous trees. The variation in sound pressure

Table 5.3. Octave-band sound pressure level for heavy trucks and automobiles for two speeds measured at 1.2 m above the ground and 15 m distance from the source (Bowlby 1998). Reprinted with permission of the Acoustical Society of America, copyright 2005

Speed (km/h)	Octave-band center frequency (Hz) 125	250	500	1,000	2,000	4,000	A-weighted sound level dB(A)
				Heavy trucks			
56	87	84.5	81.5	78	74.5	70.5	83.5
88	87.5	85	87.5	82.5	77	73.5	87.5
				Automobiles			
56	65	61	62	61	57	53	65
88	71	68	66	68	66	60	72

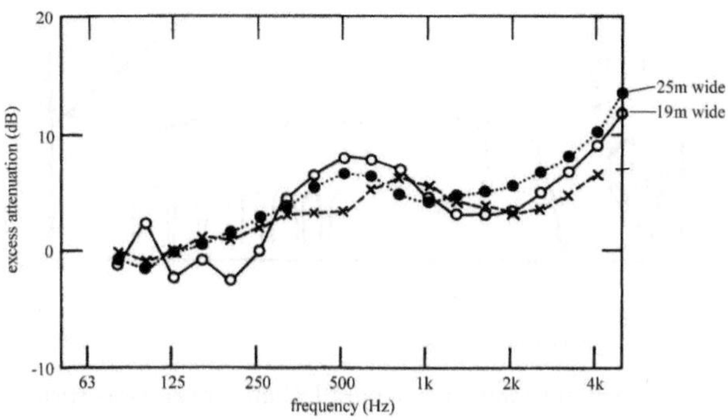

Fig. 5.3. Excess attenuation versus frequency in one-third octave bands for a belt of trees (Kragh 1981). Reprinted with permission from Elsevier. o Tree belt 19 m wide, • tree belt 25 m wide, × corn field

level versus distance clearly shows the important attenuation induced by the forest stand. The important attenuation in the forest stand due to scattering by trunks can be seen at 1.65 m (5 ft). The effect of foliage is observed at 5 m height (15 ft). Above the top of the canopies, at 14.85 m (45 ft), the attenuation is not influenced by distance in the forest stand. A strong attenuation with distance was observed with measurements between 8.25 m and 11.5 m (25 ft and 35 ft), within the canopy.

The impact of traffic noise on L_{10} and L_{95} measured in a big urban park of English style (153,500 m^2) situated in the center of a very noisy city (Athens, in Greece) showed that a reduction of 4 dB(A) was induced by the dense vegetation of shrubs and trees, at a distance of 20–40 m from the garden perimeter (Papafotiou et al. 2004). Also in Europe, the city of Geneva in Switzerland developed an outstanding policy of parks development since 1863 (Beer 1996), having today 40,000 trees, of which 85% are situated in parks and 15% along avenues and in public squares, to which must be added more than 30,000 trees on private property; and 20% of the urban surface is covered by parks. More than 350 species have been recorded, of which 50 are indigenous. Muir (1984) studied the silvicultural criteria for the selection of trees

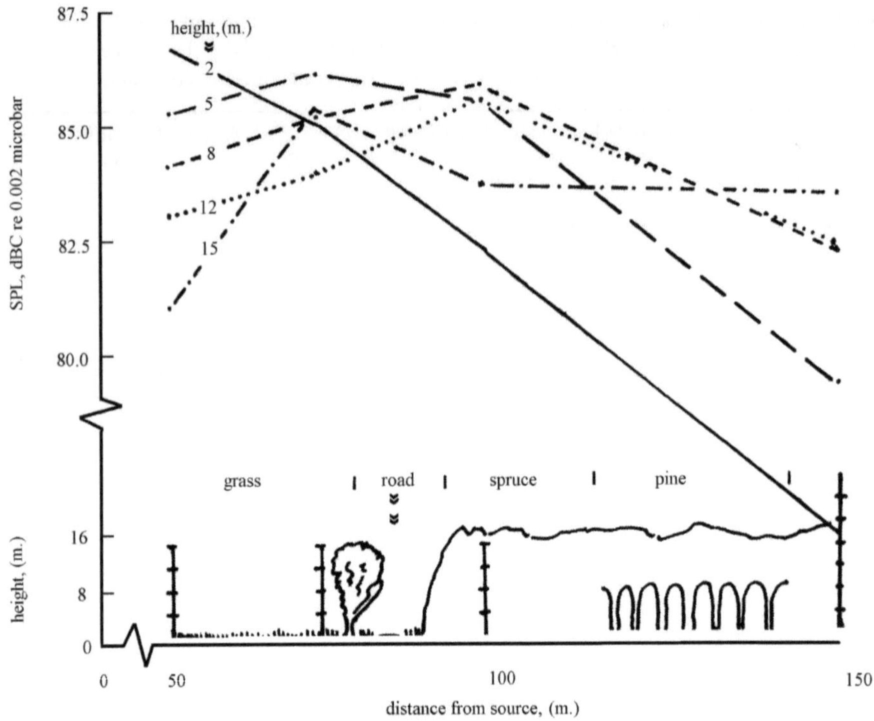

Fig. 5.4. Noise propagation over a forest stand composed of parallel bands of Norway spruce and red pine and having the edge limited by a road and a row of deciduous trees (Herrington and Brock 1977)

in urban and suburban areas, such as the growth requirements of each tree species, silvicultural background and specific features which can be evaluated for individual trees and stands. Air, water, minerals and sunlight are necessary to maintain tree growth. Parameters such as age, height, diameter at breast height, growth rate, live crown ratio, density/number of stems per unit area, defects/scars, insect damage, disease, mushrooms on the stem, etc. and general sanitary state are important for a silvicultural evaluation of trees.

Street trees have an important effect on summer micro-climate and noise abatement (Mao et al. 1993). Figure 5.5 gives an example of the arrangements of trees of different species in two streets in Nanjing City in China, in which vehicle lanes, bicycle lanes and sidewalks coexist. Depending on species, the attenuation efficiency varied between 0.1 dB/m and 0.36 dB/m (Table 5.4).

The structure and composition of street-side trees in residential areas depends on site plans, the socio-economic status of the residents and individual preferences for street-side vegetation (Zipper et al. 1991).

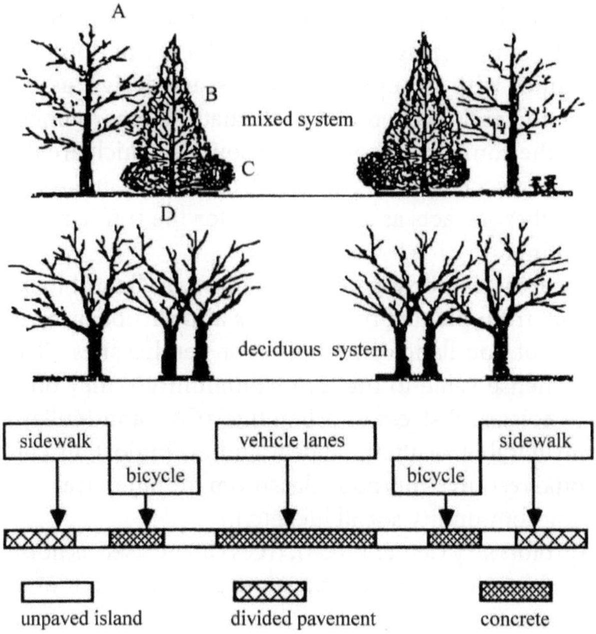

Fig. 5.5. Arrangement of trees and traffic lanes in two streets in Nanjijg, China (Mao et al. 1993). *A* = *Carya illinoensis*, *B* = *Cedrus deodara*, *Metasequoia glyptostroboides* or *Sabina chinensis*, *C* = *Pittosporum tobira* and *Euonymus japonica*, *D* = *Platanus x acerifolia*

Table 5.4. Some characteristics measured in four streets in Nanjing in China (Mao et al. 1993)

Characteristics	Streets			
	No. 1	No. 2	No. 3	No. 4
Street width (m)	40	42	28	30
Species	Deciduous	Mixed	Deciduous	Mixed
Number of tree rows	6	4	2	4
Width of green belt (m)	35	35	28	26
Canopy height (m)	4–25	4–22	4–25	4–20
Crown projection (%)	80–85	80–85	80–90	80–85
Noise attenuation dB(A)	6	4	1	8
Efficiency	0.24	0.31	0.10	0.36
Species	*P. x acerifolia*	*M. glyptostr.* *S.chinesis* *P. tobira* *C. illinoensis* *E. japonica*	*P. x acerifolia*	*M. glyptostr.* *C. deodara* *C. illinoensis*

Reduction of noise annoyance in modern urban zones can be improved by a new approach on the environmental acoustics of urban open space, proposed by Ge and Hokao (2004). The importance of urban park soundscape management has been pointed out, by exploring the relations between the soundscape of urban parks and the external environment. In the park soundscape, the main disagreement is introduced by traffic noise. Preventing the traffic noise from intruding into parks and minimizing the negative influence of noisy everyday activities can be obtained with an appropriate landscape design, as can be seen from Fig. 5.6. Protection and regeneration of the natural environment leads to the development of the components of natural sounds, which are the sounds produced by birds and insects chirping, foliage tree rustling and water flowing and jetting. The woody zone acts as a buffer area for the traffic noise, for the benefit of noise-proofing.

The woodlands in urban areas can originate from large woodland blocks which survived urbanization, from planted belts of trees and shrubs or from the spontaneous regeneration of woodland communities on derelict sites. "The urban woodlands are of immense value to the local community – they offer a temporary oasis from the noise and stress of urban life" (Cole and Mullard 1982). Also, trees have considerable aesthetic appeal and an important cultural impact as an educational resource (outdoor classroom for biological and ecological studies, improving community social life, etc.).

As noted by the USDA Agroforestry Center (2004), trees can be used as noise buffers, able to reduce noise by 5–10 dB, following some general recommendations, such as:

- plant the noise buffer close to the noise source, rather than close to the area to be protected;
- plant trees/shrubs as close together as the species will allow without being overly inhibited;
- when possible use plants with dense foliage: a diversity of tree species, with a range of foliage shapes and sizes within the noise buffer may also improve noise reduction;
- foliage of the plants should persist from the ground up: a combination of shrubs and trees may be necessary to achieve this effect;

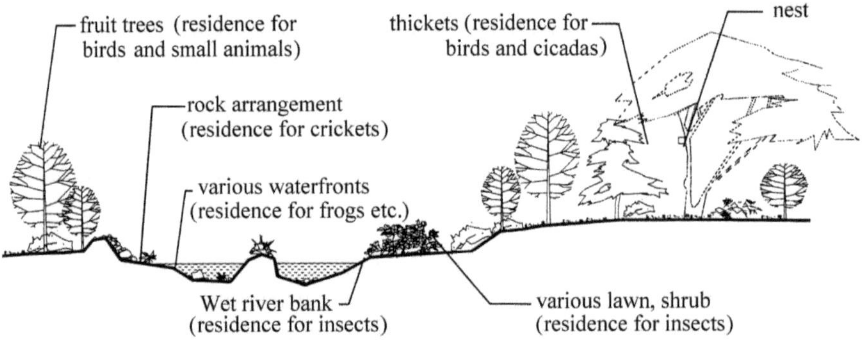

Fig. 5.6. Trees of different species and bushes for the improvement of the soundscape in urban parks (Ge and Hokao 2004)

- evergreen varieties that retain their leaves will give better year-round protection;
- when possible use taller plants. Where the use of the tall trees is restricted, use combinations of shorter shrubs and tall grass or similar soft ground cover, as opposed to harder paved surfaces.

To reduce the noise of moderate traffic in communities:

- plant belt trees 7–17 m wide along roadsides;
- plant the nearest edge of the belt within 7–17 m of the center of the nearest traffic lane;
- use 2–3 m shrubs next to the road and back-up tree rows a minimum of 4–7 m tall when mature;
- the length of the tree belt should be twice as long as the distance from the road to the recipient of the noise;
- the buffer should also extend an equal distance in both directions parallel to the road.

To reduce heavy vehicle noise in suburban or rural areas:

- plant belts of trees 20–35 m wide along roadsides;
- plant the nearest edge of the belt within 20–25 m of the center of the nearest traffic lane;
- use 2–3 m shrubs next to the road and back-up tree rows a minimum of 15 m tall at the center row;
- the length of the tree belt should be twice as long as the distance from the road to the recipient of the noise;
- the buffer should also extend an equal distance in both directions parallel to the road.

5.2 Rail Transportation Noise

Rail transportation is one of the most used systems through the world for passengers and freight within urban and suburban areas and between cities. The principal sources of noise are produced by the propulsion system of the railcars and locomotives, by the interaction between the wheels and the rail and by aerodynamics-connected phenomena (Hanson et al. 1998). The noise produced by rail transportation is expressed in terms of sound pressure levels at a standard distance from the track (30 m) and at a standard height above the ground (1.5 m).

Figure 5.7 illustrates the variation of the A-weighted sound level during 30 s corresponding to the pass-by of a locomotive-hauled passenger train at 30 m,

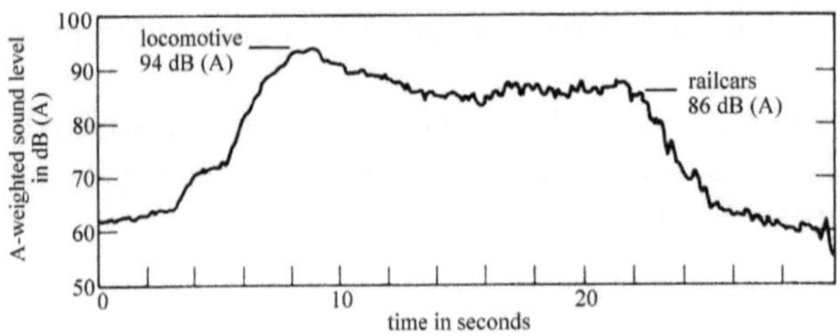

Fig. 5.7. Variation in noise level versus time (10 s) during pass-by of a locomotive-hauled passenger train travelling at 114 km/h (Hanson et al. 1998). Reprinted with permission of the Acoustical Society of America, copyright 2005

travelling at 114 km/h. The locomotive is located at 94 dB(A) and the railcars at 86 dB(A). The noise source is different for welded rail and for jointed rail.

The relationship, which predicts an increase in A-weighted sound level of 9 dB(A) per doubling of car speed, between the A-weighted sound levels and the speed of the train travelling *on continuous welded rail* is:

$$L_A = 75 + 30 \log_{10} \frac{V}{V_0} \quad \text{[in dB(A)]} \tag{5.1}$$

where V is the railcar speed (km/h) and V_0 is the reference speed (60 km/h).

When the train is travelling on *jointed rail*, the relationship between L_A and the railcar speed is:

$$L_A = 79 + 30 \log_{10} \frac{V}{V_0} \quad \text{[in dB(A)]} \tag{5.2}$$

In urban zones, trains of railcar-size passenger vehicles with rubber tires are also largely used. These trains operate at relatively low speed, 60 km/h. Because of their electrical and mechanical equipment, the noise levels produced by these types of trains are similar to each other.

Rail systems generate groundborne vibrations which are important and depend on the resonance frequencies of the train suspension systems and the smoothness of the wheels and rails.

The main desiderata of modern acoustic research are related to the reduction of noise annoyance from 91 dB(A) to 83 dB(A) for high-speed trains and a reduction between 20 dB(A) and 10 dB(A) for freight within urban and suburban areas (Gautier et al. 2004).

Railway noise, which is short and abrupt, recurring only after a long period, is different from road traffic noise, which is mostly continuous and has short periods of discontinuity (Raimbault et al. 2004). Human subjects recognize railway noise as related to specific time-patterns, characterized by the identification of the noise source, whereas road traffic noise is related to a description related to the space generating the noise. Environmental factors and visibility can enhance or mask noise loudness (Mulligan et al. 1987; Watts et al. 1999). At the same time, the variability in the perception of noise loudness is related to the individual sensibility of subjects (Mace et al. 1999).

The influence of meteorological conditions on the noise attenuation of rail transportation is mentioned in the French standard AFNOR XP S31-133 (2001).

Rail noise reduction by the belt of trees was studied by Kragh (1979). The experimental arrangement is shown in Fig. 5.8 for two distances between the microphones, namely 68 m (case 1) and 40 m (case 2). In the first case (Fig. 5.8a), the rails were 0 to 0.5 m below terrain. The tree belt was of a finite length of 400 m, composed of 50-year-old birches and elms mixed with lower 15-year-old beeches and confers. The position of the microphones is indicated in the figure. The track consisted of welded rails on rubber plates on wooden sleepers in stone ballast. In the second case (Fig. 5.8b), the belt was 1,200 m long and

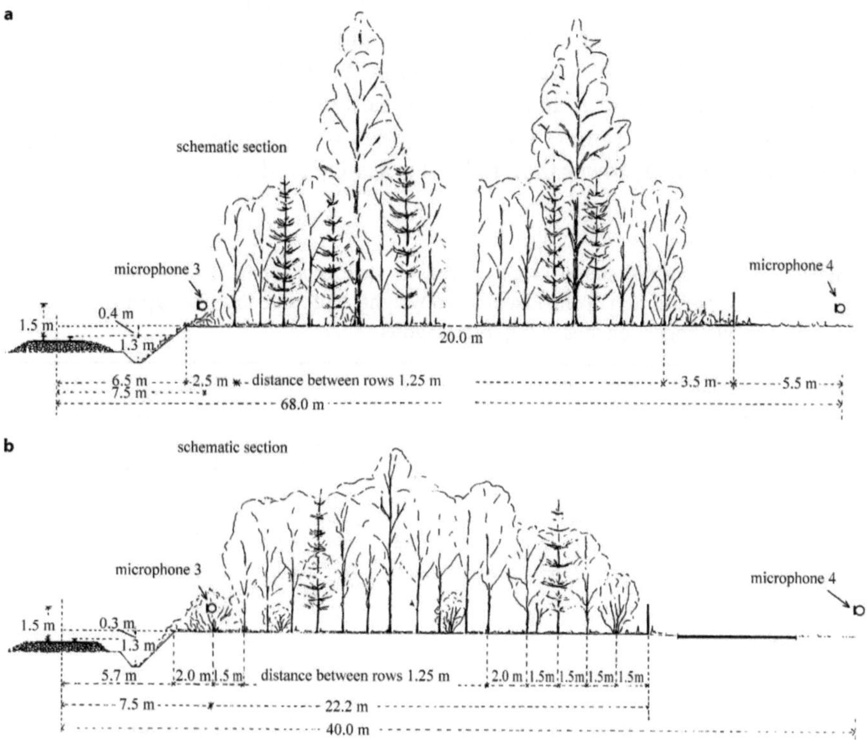

Fig. 5.8. Experimental arrangements for sound level measurement from passing trains and attenuation measurements through belts of trees and bushes. Reprinted from Kragh (1979), Copyright 2005, with permission from Elsevier. **a** Belt profile and measurements for 68 m distance between the microphones, **b** belt profile and measurements for 40 m distance between the microphones

25 m wide, composed of 20-year-old oaks mixed with hornbeams, poplars and silver firs, and bushes. The track consisted of 30-m lengths of jointed rails, directly on wooden sleepers in stone ballast.

Noise from the same train passing by was recorded in two positions, at the reference section of the track and at the belt positions. Noise from 15 pass-bys was recorded during 8 h at each site. In the laboratory, for each microphone position, the $L_{A_{eq}/60s}$ was determined for each microphone. The A-weighted sound pressure levels were determined. The difference was calculated between $\Delta L_{A_{eq}}$ and $L_{A_{eq}}$ at positions 1 and 2 and positions 3 and 4. The difference $\Delta(\Delta L_{A_{eq}})$ was calculated at shelter belt positions 3 and 4 above the reference attenuation at positions 1 and 2 (Table 5.5). As can be seen from this table, the main value of $\Delta(\Delta L_{A_{eq}})$ at site 1 was 9 dB(A), with a standard deviation of 1.4 dB(A) and 6 dB(A) at site 2.

Wheel/rail noise had a major contribution to the overall noise level.

A definitive conclusion is difficult to draw from this experiment, because firstly the source of noise was not identical at sites 1 and 2 and secondly the site configurations had minor differences. In any case, it can be stated that the attenuation effect of belt vegetation is combined with terrain configurations. In the first case, behind a dense belt, 15 years old, 50 m wide, composed of beeches and various conifers planted between older birches and elms, noise levels were 8–9 dB(A) lower than in level grass-covered country. In the second

Table 5.5. Acoustic characteristics measured in two experimental sites (Kragh 1979). Reprinted with permission from Elsevier, copyright 2005. Locomotive type: diesel electric; cars: p = passenger coaches, g = goods wagons

Site	Loco. type	Cars	Speed	$L_{A_{eq}/60s}$ [dB(A)]				$\Delta L_{A_{eq}}$ [dB(A)]		$\Delta(\Delta L_{A_{eq}})$ [dB(A)]
		(no./type)	(km/h)	Pos. 1 7.5 m	Pos. 2 55 m	Pos. 3 7.5 m	Pos. 4 68 m	1 & 2	1 & 3	3,4 & 1,2
No. 1	MY	4p	105	85.4	74.2	86.4	64.5	11.2	21.9	10.7
(Fig. 5.8a)	MV	10g	85	81.8	70.0	82.4	60.2	11.8	22.2	10.4
Average site 1 for 13 trains	–	–	–	–	–	–	–	13.4	22.4	9.1
Standard deviation	–	–	–	–	–	–	–	1.7	0.4	1.4
No. 2	MY	4p	108	88.3	78.7	92.4	74.3	9.6	18.1	8.5
(Fig. 5.8b)	MY	12g	79	89.3	75.6	90.4	73.1	13.7	17.3	3.6
Average site 2 for 15 trains	–	–	–	–	–	–	–	12.4	18.3	15.9
Standard deviation	–	–	–	–	–	–	–	1.7	0.7	1.8

case, behind a dense belt, 10–20 years old, 25 m wide, composed of oaks, hornbeams, poplars, silver firs and bushes, noise levels were 6–7 dB(A) lower than in the level grass-covered country. Without doubt, it can be concluded that the belts of trees are effective in railway noise attenuation.

5.3 Aircraft Noise

Aircraft and helicopters generate annoying noise in urban, suburban and natural recreational environments, which interferes with the aesthetic quality of the landscape (Raney and Cawthorn 1998). For example, the noise level measured in Grand Canyon National Park in the United States is 76 dB (A) for aircraft and 40 dB(A) for helicopters, as noted by Mace et al. (1999).

Aircraft noise is increasingly seen as a constraint on the future growth of aviation traffic. The noise generated by aircraft components is mainly produced

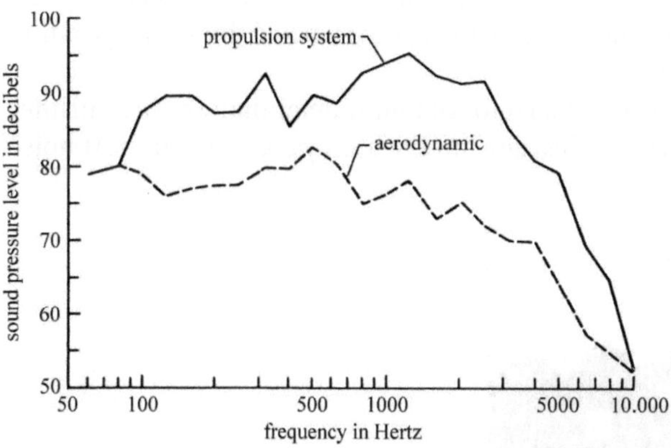

Fig. 5.9. Sound pressure level versus frequency for the aerodynamic noise of a turbofan aircraft compared with propulsion system noise landing approach at an altitude of 150 m and speed of about 125 knots (Raney and Cawthorn 1998). Reprinted with permission of the Acoustical Society of America, copyright 2005

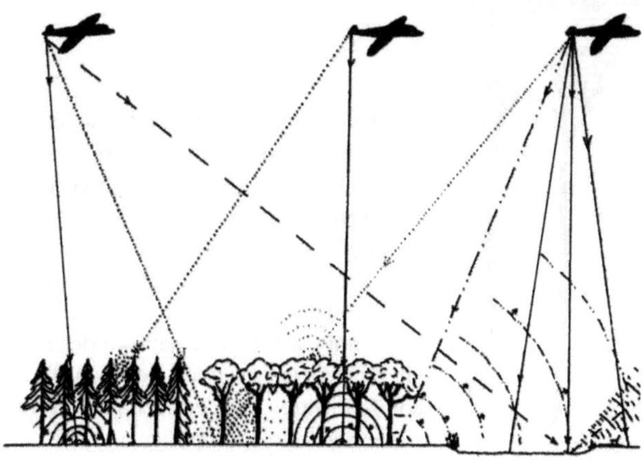

Fig. 5.10. Aircraft noise propagation in coniferous and deciduous forest stands (von Wendorff 1974)

by the engine's rotating blades, by the air expelled from the jet exhaust, by the airframe structure as it flies through the air, by the strong local airflow of devices and surfaces deployed during aircraft take-off or landing, etc. Figure 5.9 shows the variation in the sound pressure level versus frequency for the propulsion system and for aerodynamic noise. The maximum sound pressure level (95 dB) was observed around 1 kHz.

The mechanism of aircraft noise propagation through forest is demonstrated in Fig. 5.10, which shows the aircraft displacement in three successive positions. The noise radiated from the aircraft propagates through the atmosphere and interacts with the forest stand and the ground. Figure 5.11 proposes a qualitative model for noise propagation phenomena in a forest stand. Four typical zones inside the forest stand favorable, or not, to noise propagation were defined by von Wendorff (1974) as follows: zone A, which corresponds to free noise propagation in atmosphere; zone B, which corresponds to the canopy zone, mainly absorbing the noise; zone C, or a shadow zone in which the noise is mostly reflected; zone D, or a resonance zone situated under the canopy and acting as a wave guide. In the figure, the characteristic free space between trees noted 'a' for coniferous forest stand or 'd' for deciduous trees acts as a noise amplifier; and, for this reason, the author labeled this corresponding amplifying effect as the 'trumpet effect'.

Today's best airliners are about a factor of four quieter than the first airliner introduced more than four decades ago, but the annoyance with aircraft noise

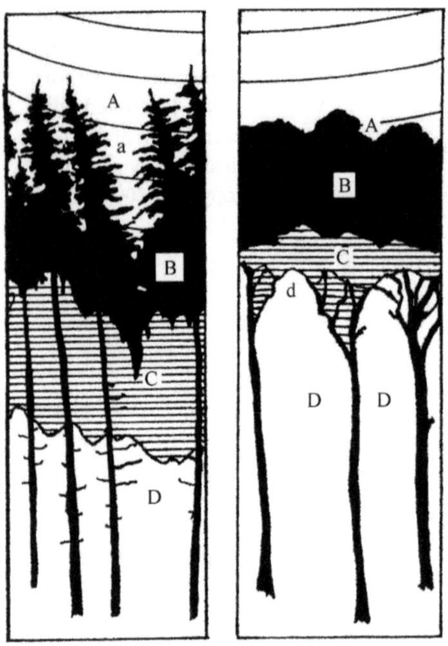

Fig. 5.11. Typical zones inside the forest stand favourable or not to noise propagation (von Wendorff 1974). *A* Free noise propagation in atmosphere, *B* zone of canopy, absorbing the noise, *C* shadow zone in which the noise propagation is reflected, *D* resonance zone under the canopy acting as a wave guide. The characteristic free space between the coniferous trees (*a*), or deciduous trees (*d*) acts as a noise amplifier

in recreational areas and at home is very much criticized because of increasing aircraft traffic and airport noise, since noise has been recognized as a serious environmental pollutant. An annoyance weighting factor has been proposed to express annoyance in social surveys by residential populations living near airports (von Gierke et al. 1998). People more highly annoyed at home tended to be more annoyed than others while in recreational areas (Krog and Engdahl 2004). Noise sensitivity in response to aircraft environmental noise is an independent predictor of annoyance, which statistically can explain up to 26% of the variance (van Kamp et al. 2004).

Forest stands planted near airports can be a good solution for the reduction of the noise annoyance produced by the continuous growth of aviation traffic and can contribute to the well-being of the traveling public and the communities surrounding airports.

5.4 Summary

Outdoors, a description of the noise environment in the audible range is provided by the A-frequency weighted day-and-night average sound level. The environmental noise impact depends on the total energy received at the observation point, the rate of occurrence of noise events and the magnitude of noisier single events. The reduction in noise level requires the following three main steps: evaluation of the noise environment under existing conditions, determination of the acceptable noise level and determination of the difference between the two previous steps. In urban communities, the sources of noise are numerous and may include noise produced by highways, rail and aircraft transportation. The techniques for noise control are related to control at the source, at the receiver and at the transmission path.

The main outdoor sources of noise in modern life are generated by traffic on streets or highways, by rail transportation and by aircraft.

Models of different complexity have been used for traffic noise prediction since 1950. A unification of noise calculation algorithms in Europe was realized with the standard ISO 9613-2 (1996). The propagation of highway noise over a forest stand expressed by the variation in sound pressure level versus distance has clearly shown the important attenuation produced by forest stands. In urban areas, trees can be used as noise buffers, able to reduce noise by 5–10 dB, if some general recommendations are respected (plant trees near the noise source, plant trees/shrubs with dense foliage as close as possible, plant belt trees of 7–17 m wide, etc.).

Rail transportation is one of the most used systems throughout the world for passengers and freight within urban and suburban areas. The noise is produced by the propulsion system of the railcars and locomotives, by the interaction between the wheels and rail and by aerodynamics-connected phenomena. Rail systems generate ground-borne vibrations which are important and depend on the resonance frequencies of the train suspension systems and the smoothness of the wheels and rails. The attenuation effect of belt vegetation is combined with the terrain configuration.

Aircraft and helicopters generate annoying noise in urban, suburban and natural recreational environments, which interferes with the aesthetic quality of the landscape. The noise radiated from the aircraft propagates through the atmosphere and interacts with the forest stand and the ground. Forest stands planted near airports can be a good solution for reducing the noise annoyance produced by the continuous growth of aviation traffic.

6 Noise Abatement and Dwellings

Consideration of noise and sound elements in urban open space is absolutely important today in the daily life of citizens. The design of a comfortable environment must pay attention to the "*soundscape*", as defined by the Canadian musician Murray Schafre at the end of 1960s, which should complete the "*landscape*" design, which refers to a visual scenario of the environment. In soundscape, the sounds produced by human beings, different kind of traffic noise, birds, animals, waterfalls or streams, bells for traditional events, broadcasting, background music and other natural, artificial or social sounds, etc., should be integrated into one united sound environment, related to the different personal feelings of humans in various circumstances. The meanings of soundscape are social, historical, cultural and environmental (Ge and Hokao 2004). The concept of soundscape was promoted by international organizations and congresses (Inter-Noise) and has been applied to practice (sound maps) for urban planning, environmental, architectural and equipment design. Urban parks are typical subjects in which functions, such as sports, relaxation, cultural and social events, sightseeing, etc., are considered as environmental elements. Recording the soundscape components helps to catch the main sound components and their structure on a scale in terms of "preference" or "congruence" and to serve as a base for future improvements. A positive impression on the urban soundscape is produced by large vegetation areas, belts of trees, public gardens and parks.

6.1 Urban Area

The interest of acousticians for noise abatement in urban areas with vegetation, shrubs or a belt of trees is quite old (Embelton 1963; Aylor 1972a, b; Cook and van Haverbeke 1971, 1972, 1977; Haupt 1973, 1974; Herrington 1974; Kellomäki et al. 1976; Reethof and Heisler 1976; Ishii 1994). The significance of height, density, width and length of tree belts for noise reduction was underlined. Recent studies (Fang and Ling 2003) included a new parameter, namely the visibility (see Chap. 2).

Figure 6.1a, b suggests a possible disposition of trees around a house for maximum noise reduction all year round, together with an improvement in aesthetics and air quality, in a residential urban area.

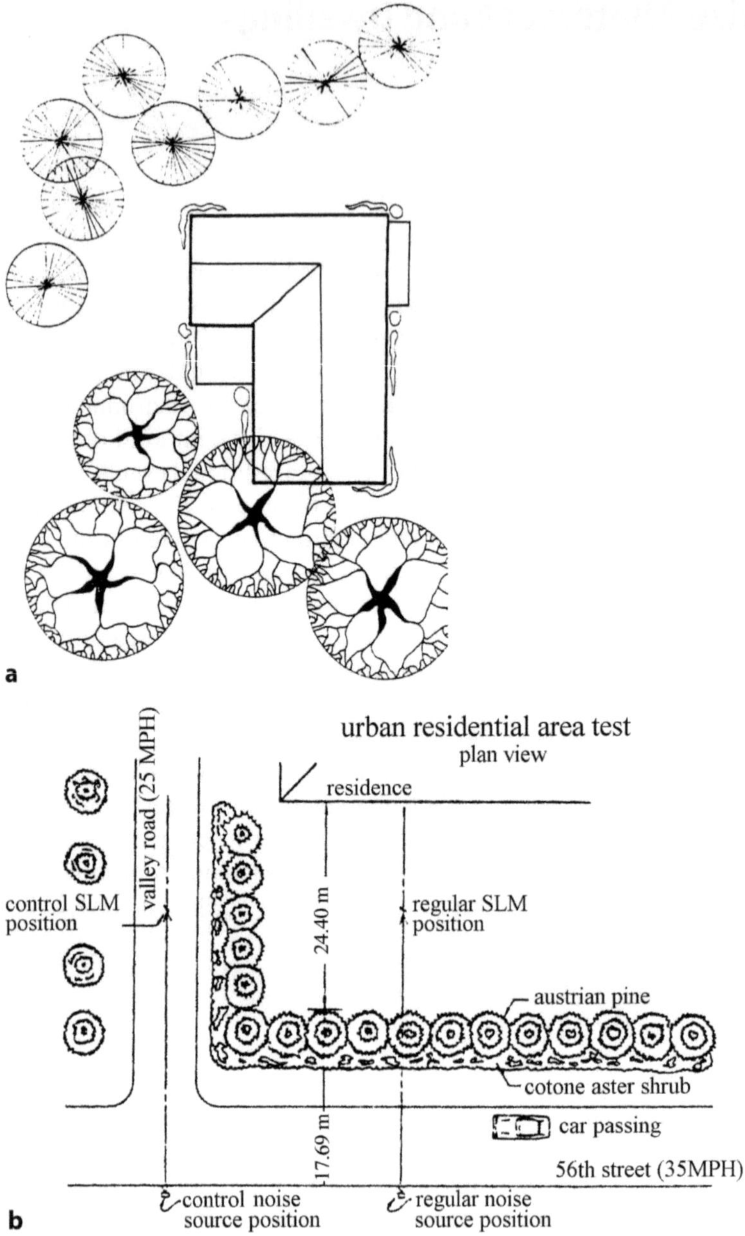

Fig. 6.1. Trees around a house for maximum noise reduction as used in urban residential area in USA: **a** all year round, with high, big coniferous and decidous trees (Heisler 1974), **b** plan view of dense trees and shrubs (Cook and van Haverbeke 1972); N.B 35 mph = 56 km/h car passing speed

A typical tree belt structure (Fig. 6.2), composed of coniferous and deciduous species, reduced the sound level produced by urban buses, autos and trucks, by about 40 dB for a distance of 150 m (Fig. 6.3).

Kellomäki et al. (1976) studied the characteristics of coniferous tree stands suitable for noise reduction in urban areas in countries of northern Europe. Attenuation and excess attenuation by coniferous stands of different ages compared to a cut area are given in Table 6.1. The early succession stage of a stand gives better attenuation than mature stands. In the case of a spruce stand,

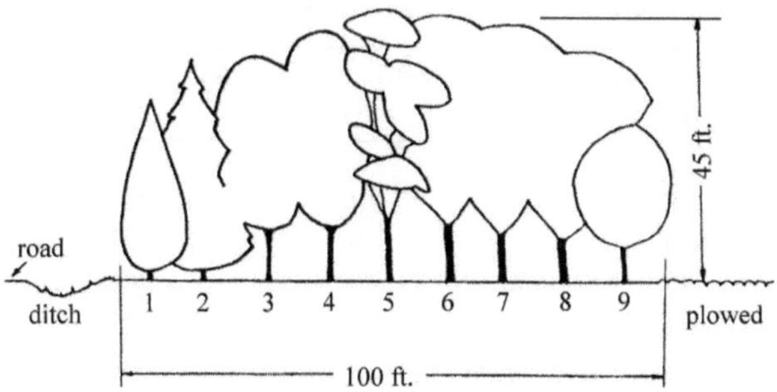

Fig. 6.2. Possible configuration of the typical structure of a tree belt in an urban area, composed of the following species: 1 eastern red cedar, 2 ponderosa pine, 3 green ash, 4 hackberry, 5 honey locust, 6 Siberian elm, 7 Siberian elm, 8 American elm, 9 mulberry (Cook and van Haverbeke 1972); 100 ft = 30.5 m; 45 ft = 13.75 m

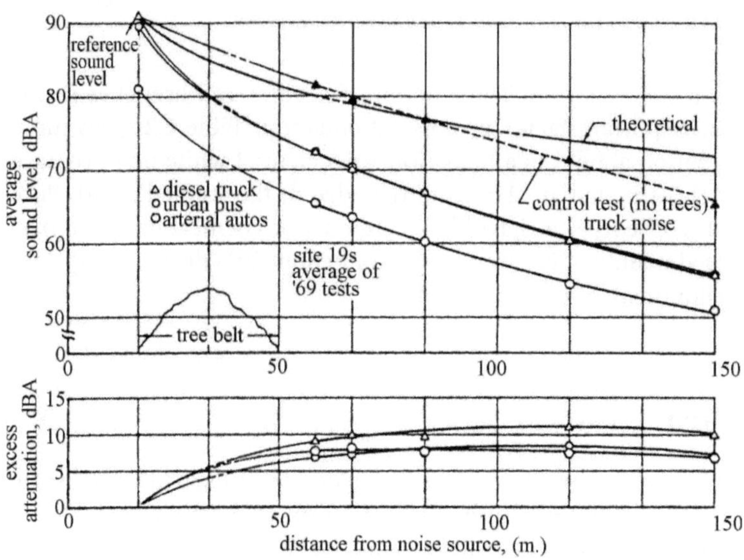

Fig. 6.3. Noise attenuation produced by a belt of trees. Relationship between sound pressure level and distance (Cook and van Haverbeke 1972)

Table 6.1. Attenuation of stands compared to clear area (Kellomäki et al. 1976)

Species	Attenuation of stands compared to clear area			
	Seedling stand	Middle-aged stand	Mature stand	Mean value
Pine	0.740	0.583	0.825	0.716
Spruce	0.307	0.346	0.510	0.388
Mixed	0.340	0.290	0.518	0.383
Mean value	0.462	0.406	0.617	0.495

Table 6.2. Physical and acoustic characteristics of stands (Kellomäki et al. 1976)

Stand	Tree density Number/ha	Mean height (m)	Volume (m^3)	Dominant tree (%)	Self-pruned tree (%)
Pine					
Seedling	500	16	128	14	55
Middle-aged	700	15	134	8	40
Mature	1,000	15	108	10	38
Spruce					
Seedling	1,633	5	11	9	6
Middle-aged	900	19	208	3	22
Mature	600	18	287	2	24
Mixed					
Seedling	2,433	8	71	4	13
Middle-aged	1,133	10	195	1	19
Mature	600	22	227	1	36

60% attenuation was reported. The amount of needles and branches seems to be the most important factor in noise attenuation. Table 6.2 gives more detailed information about the attenuation coefficient of white noise (octave bands 20–100 Hz) situated at 12 m from the edge of the studied stand. Pure stands composed of Scots pines have higher attenuation coefficients than mixed stands composed mainly of Norway spruce and about 20% deciduous species with birch predominance.

6.2 Suburban Area

One of the main sources of discomfort in residential suburban areas is the traffic noise produced by the continuous expansion of the highway systems. The reduction of this traffic noise can be achieved in two ways:

- First, by lowering the speed limit of the vehicles and improving engine muffling. (A large truck exceeds 100 dB.) This approach is very limited.

- Second, by creating trees belts and shrubs and solid barriers between the noise sources and dwellings. The barriers can be walls up to 2 m in wood, masonry, earth dikes or small natural hills.

In general, it can be said that the amount of noise reduction measured at the receiver is dependent on the in situ configuration and is mainly determined by the level at the noise source and by personal sensitivity. An acceptable level is 70 dB for daytime activity and 50 dB for evening time. A reduction of 10 dB produces a sensation of noise cut by about half.

In what follows we propose an analysis of practical situations, as presented by Cook for the effect of a noise source produced by traffic at 35 mph (56 km/h) on a highway (automobiles and trucks) at three sites noted A, B and C. Site A has low, dense cotoneaster shrubs backed by tall ponderosa pines. Site B has medium-height planting, a woven board fence and a downward sloping ground profile from the street to the residence. Site C has a tall evergreen hedge, a brick wall and an upward sloping ground profile from the street to the residence. A schematic presentation of the experimental configuration for noise level measurement is shown in Fig. 6.4. The average sound level and the relative attenuation function of the distance from the source are shown in Figs. 6.5, 6.6, 6.7. A noise reduction ranging from 45 dB to 10 dB was measured. Figure 6.8 gives a comparison of the noise reduction produced by the trees, the wall and the tree–wall combination. As expected, the highest reduction was obtained by the tree–wall combination, for a wall 1.8 m in height. Figure 6.9 shows the effect of wall placement between the source and the receiver. The maximum attenuation seems to be obtained at 10 m distance from the wall.

The analysis of different configurations allows one to say that, in residential suburban areas, plant material can reduce noise levels by as much as 8 dB, when the residence is at least 25 m from the centerline of the roadway. Coniferous

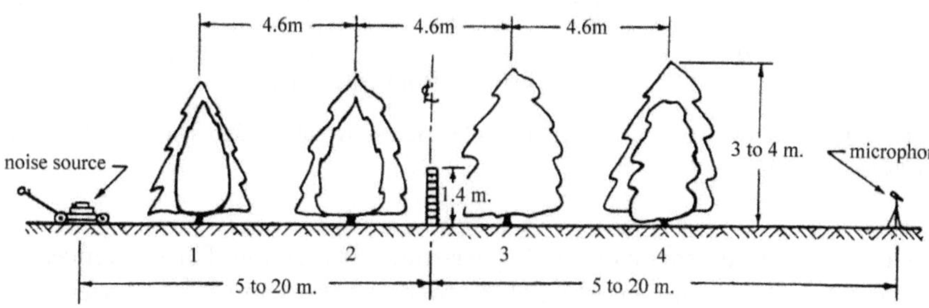

Fig. 6.4. Schematic diagram of experimental configuration for noise measurement with trees and wall. Tree spacing in all rows is 2 m (Cook and van Haverbeke 1977). *1* Austrian pine alternated with rocky mountains juniper, *2* ponderosa pine alternated with eastern red cedar, *3* Scots pine, *4* Austrian pine alternated with oriental arborvitae

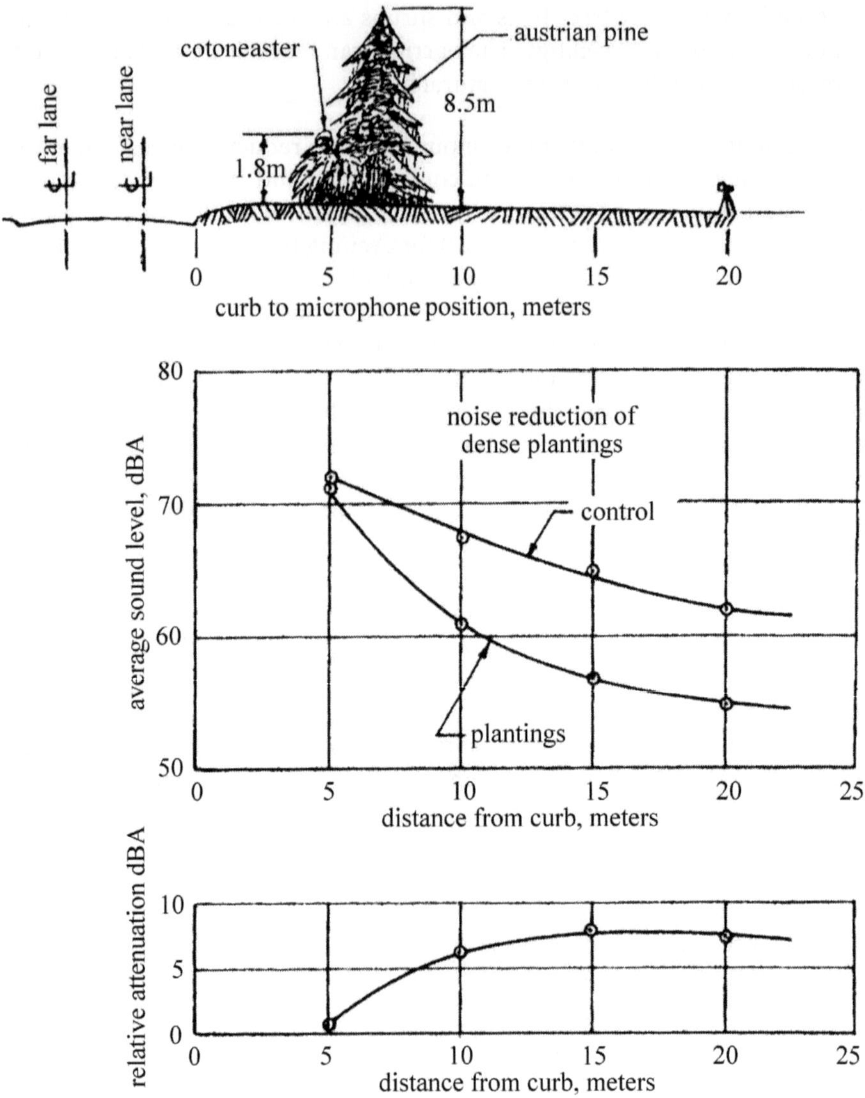

Fig. 6.5. Site A has low, dense cotoneaster shrubs backed by tall ponderosa pine (Cook and van Haverbeke 1972)

species or evergreen shrubs provide year-round protection. For residences at less than 20 m from the centerline of the roadway, trees and solid barriers are required.

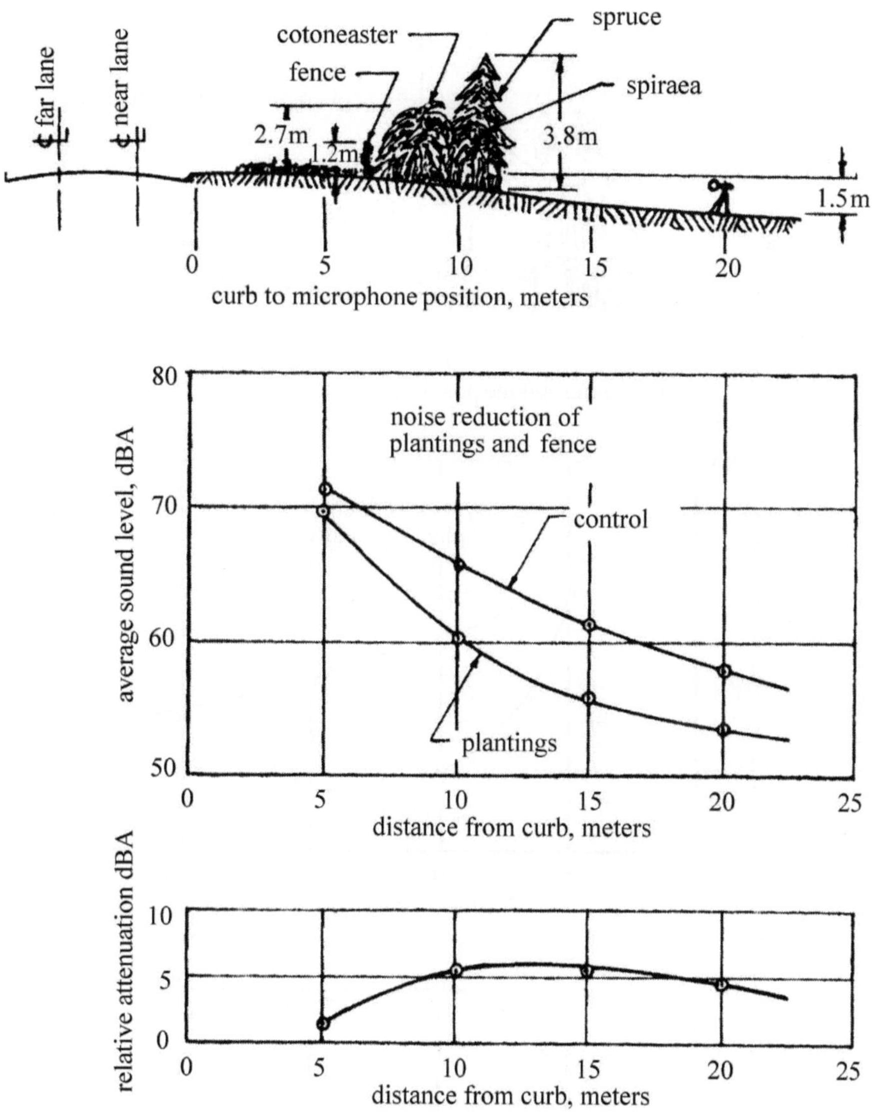

Fig. 6.6. Site B has medium-height planting, a woven board fence and a downward sloping ground profile from the street to the residence (Cook and van Haverbeke 1972)

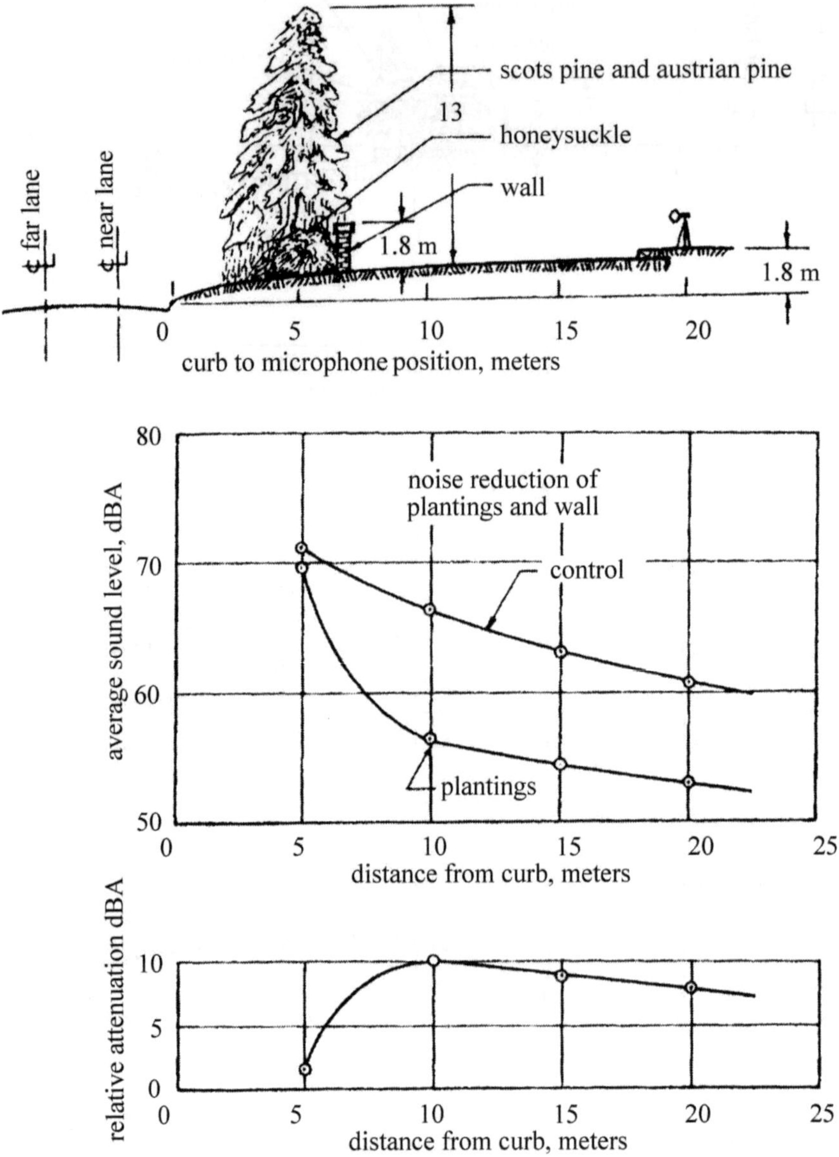

Fig. 6.7. Site C has a tall evergreen hedge, brick wall and an upward sloping ground profile from the street to the residence (Cook and van Haverbeke 1972)

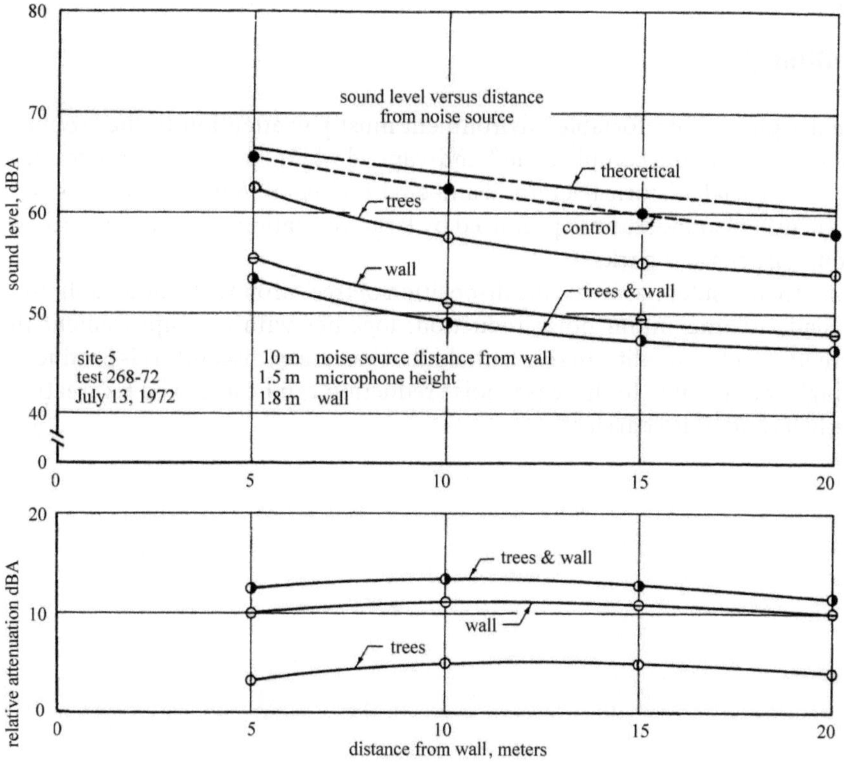

Fig. 6.8. The influence of different surfaces – wall alone, trees alone, trees and wall – on noise reduction (Cook and van Haverbeke 1972)

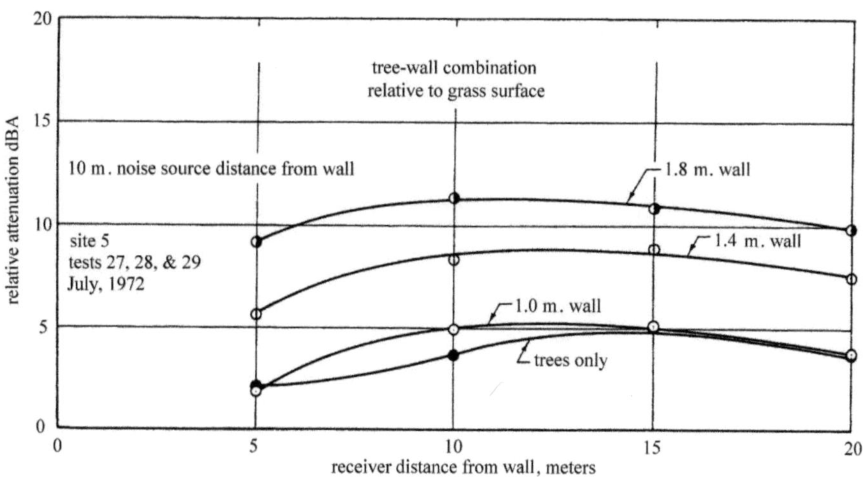

Fig. 6.9. The effect of the position of the wall between the source and the receiver (Cook and van Haverbeke 1972)

6.3 Summary

The design of a comfortable environment must pay attention to the "soundscape" which should complete the "landscape" design. The meanings of soundscape are social, historical, cultural and environmental. A positive impression on the urban soundscape is produced by large vegetation areas, belts of trees, public gardens and parks.

In urban residential areas, the disposition of trees around the houses should be made for maximum noise reduction, together with an improvement in aesthetics and air quality. In residential suburban areas, discomfort is produced by highway systems. In this case, noise reduction can be achieved by creating tree belts and noise barriers.

7 Noise, Birds and Insects in Urban Forest Environment

In the urban forest environment two major groups coexist: the birds and the insects.

In the biological world of animals, communication by sound plays an important part. The interest on the influence of habitat on acoustic communication by animals is relatively new; and probably the first article was published by Chappuis (1971). Since then, this field has undergone important development, as demonstrated by published books (Kroodsma et al. 1982; Bradbury and Vehrencamp 1998) and other reference articles, from which because of space limitation we cite only two (Padgham 2004; Slabbekoorn 2004).

Sources of ambient noise in the urban forest habitat are produced by abiotic and biotic noise sources and also by anthropogenic noises, e.g. low-frequency rumbles and roars from cars, aircraft, etc. Rustling leaves and twigs have a spectrum over a wide range of frequencies, while air passing over substrates produces low-frequency sounds. The impact of wind depends on the openness of the vegetation. The spectral composition of sound produced by birds, insects, etc. is habitat-specific (Slabbekoorn 2004).

The process of communication in the natural environment involves two individuals, a sender (or acoustical source) and a receiver. The acoustic signal is transmitted through the environment and is detected by the receiver. Both sender and receiver benefit from information exchange by acoustic signals (Fig. 7.1). Birds have complex interactions and communicate information

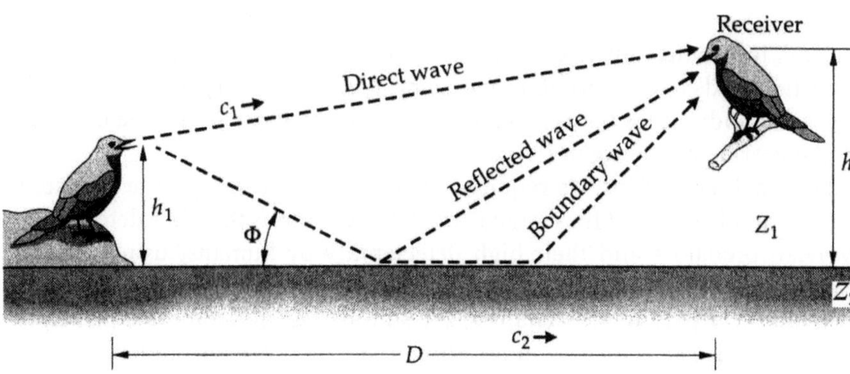

Fig. 7.1. Acoustical communications between two birds (Bradbury and Vehrencamp 1998)

about their identity (species, sex, group membership, individual identity) and mood (such as dominance, fear, agressivity), which is essential to maintain the species' dispersed spatial patterns. Bird vocalizations are modulated in a complex way during the production of acoustical signals and can be represented as a spectrogram in time and frequency domain (e.g. a graphic representation as sonograms for two-dimensional spectrograms). Long-distance transmission of acoustical signals is a crucial factor for animal communication. The distance at which the signal to noise ratio is sufficient for conspecifics to detect and recognize the signal is also an important parameter. The spectral profile of ambient noise is habitat-specific. It was observed that the song characteristics of individual birds are corrected to local noise conditions. The environmental noise alters the spectral composition of the propagated sound by adding new frequency components and sometimes new energy to existing components. The major sources of low-frequency noise in air are wind and air turbulence passing over vegetation. For a wind speed of 1 m/s, typical levels for frequencies under the 200 Hz range are between 20 dB and 30 dB SPL. For higher speed (8 m/s) 60–70 dB were measured. The sources of high-frequency noise (> 4 kHz) in terrestrial habitats are chorusing insects, producing 40–50 dB. The frequency bands of relatively quiet sounds, between wind and insect sounds, range from 1 kHz to 4 kHz, with 10–20 dB SPL.

As noted by Bradbury and Vehrencamp (1998), bird sound production involves three successive steps: the production of vibrations, their modification to match biological functions and finally transmission into the environment medium. The difficulty in producing vibrations and coupling them to the medium explains the fact why not many animal taxa use sound communication. For most bird ears, the frequency spectrum provides the most important information. During sound propagation through the environment, the spectrum can be perturbed by global attenuation, loss of pattern and additional noise.

The frequency of vocalization depends on body size, as can be seen from Fig. 7.2. It is also accepted that the larger the wavelength relative to the size of the bird, the lower the intensity of the emitted sound. For everyday life, a practical solution to produce intense sounds and avoid losses due to attenuation and distortion leads to the optimum frequency. For the majority of song birds, for which the body size is between several centimeters and < 0.5 m, the optimum frequency range is between 1 kHz and 6 kHz.

The small size of insects requires an optimal frequency in the ultrasonic range. The advantage of these signals, which are short, is their low detectability by avian predators and their high detectability by humans, using acoustic emission transducers.

In the first part of this chapter, we summarize the common principles involved in bird acoustic communication in the natural environment.

The second part of this chapter is devoted to the detection of termite infestations in urban trees.

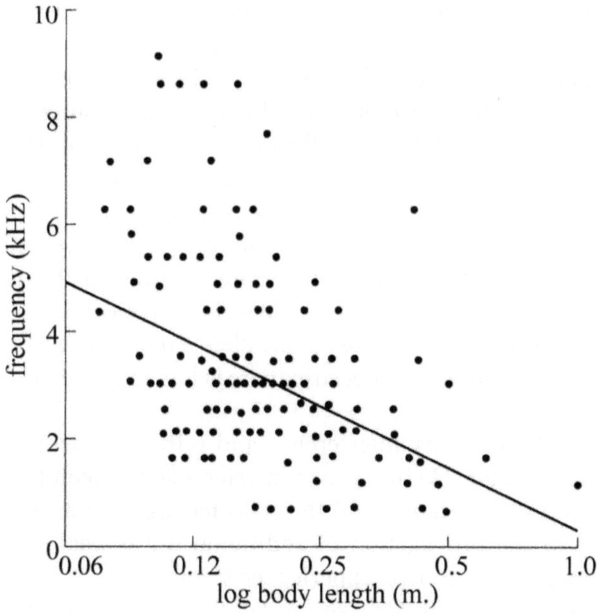

Fig. 7.2. Fundamental frequency and body size for birds (Bradbury and Vehrencamp 1998)

7.1 Bird Acoustic Communication in Forest Environment

In the natural environment, bird song is a typical communication over a distance of 50–200 m, in frequencies between 1 MHz and 8 MHz (Wiley and Richards 1982). The threshold for hearing in passerines rises steeply below 2–3 kHz and the frequencies in long-range songs are the mirror image of the hearing threshold.

In the terrestrial environment, noise distribution tends to be bimodally distributed over frequency, with a quiet band between 1 kHz and 4 kHz. Lower-frequency noise is mainly produced by winds. The sound pressure level of the overall noise due to nonhuman sources is between 45 dB and 55 dB.

As we have seen previously (Chap. 4), the attenuation of sound in the atmosphere depends on spherical spreading, atmospheric absorption, scattering and boundary interference. Attenuation is frequency-dependent; and consequently some frequencies are optimal for long-distance communication determining "sound windows", which are dependent on the location of the sender and receiver with respect to the ground. The effect of air temperature and relative humidity is important. For a temperature greater than 5 °C and relative humidity variation between 50% and 90%, a reduction in attenuation is expected by the absorption of about 1 dB/m for frequencies below 2 kHz. For current temperatures in spring (15 °C) and 4 kHz frequency, the effect of relative humidity is more important. Following the diurnal cycle in a natural environment, the temperature rises, the relative humidity decreases and the atmospheric absorption varies continuously.

Scattering in the natural environment is also frequency-dependent and is produced by vegetation and by variations of sound velocity in air because of temperature and humidity variations. Near a small source such as a bird, the sound often radiates in a beam. Deflection of the beam energy can reduce the intensity of bird song. We have seen in Chap. 4 that attenuation by scattering in forest can reach 10 dB over 100 m for frequencies above 1 kHz. Between 2 kHz and 11 kHz, foliage also increases attenuation by about 10 dB over 100 m (Marten and Marler 1977; Marten et al. 1977). In the bird song frequency range, the scattering produced by wind and atmospheric turbulence induces severe attenuation. Birds often sing in the morning because there is less atmospheric turbulence.

In the natural environment, boundary interference and refraction, which are also frequency-dependent, occur: (1) because of interference between the direct wave from the source to the receiver and the reflected wave from the boundary and (2) because of the propagation of additional waves near the ground (Fig. 7.3). The typical refraction effect experienced by birds is produced by variations in temperature and distance. On a sunny day, the ground is warm and the peak of sound velocity is at the surface of the ground. The air near the surface has a higher sound velocity than the air at greater heights. The acoustic signals are refracted up and away from the ground, generating a shadow zone at some distance from the source. In the canopy during the day, the warmest air occurs at some height above the ground. Vocalizations emitted below the warm air zone are refracted back down and travel a long distance. A shadow acoustic zone above the warm layer exists in this case.

Destructive interference primarily affects bands of low frequency (0.5–1.0 MHz) for propagation about 1 m above the ground over 100 m. Between 1 kHz and 2 kHz, the attenuation is much less pronounced, since the scattering

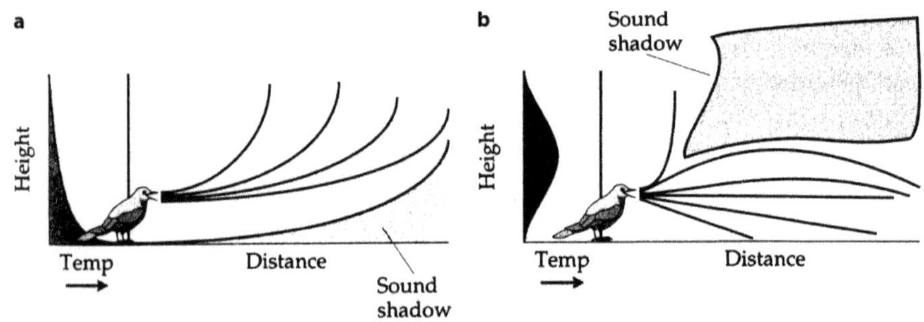

Fig. 7.3. Refraction effects as a function of temperature and distance (Bradbury and Vehrencamp 1998)

of sound destroys the coherence of direct and reflected waves. Aside from near the ground, the patterns of attenuation for bird long-range communication in all habitats favor a low frequency. The frequency-modulated tone of bird song avoids the effect of reverberation and amplitude fluctuation on the emitted acoustic signals. Degradation of the acoustic signal encodes information about the distance to the sender.

Temperature and wind greatly affect the refraction of bird song transmission. Finding and using a sound channel and avoiding the sound shadow zone will sometimes increase the pattern distortion. The height at which the sound is emitted between the canopy and the ground has a major effect on its propagation, by producing a sound channel.

Acoustic signals emitted above the canopy layer may lead to increasing attenuation. The rate of exponential reverberation decay depends upon frequency. Increasing the source height is effective for reducing reverberation.

7.2 Detection of Termite Infestation in Urban Trees

Detection of the presence of termites in urban areas is a major challenge for many town administrations all over the world, in tropical and temperate zones. Acoustic methods (namely acoustic emission) for the detection of the presence of termites in trees and wood are a valuable alternative to traditional visual inspection (Bucur 2005).

The operating frequency range of the acoustic method (40 kHz or more) guarantees the successful application of this method. Termite signals are pulses of short duration and high frequencies when compared with the background noise in urban zones. The pulse rate is proportional to the number and spatial distribution of the insects and can vary widely among sites and zones in a single tree. The number of termites depends on the species, temperature, age and physiological status of individuals.

The spectra of signals recorded for different termite species and urban background noise are given in Fig. 7.4. A relative scale was used for the graphic representation of vibration levels ranging between 0.1 kHz and 4.0 kHz.

Table 7.1 gives data reported by Mankin et al. (2002) on the mean activity rate of *Coptotermes formosanus* and *Reticulitermes flavipes* measured on standing trees. In this experiment, the minimum measured value reported was 4.7 pulses/s. The minimum number of detected insects was 55 for a density of 0.06 termites/cm and per gram of wood.

As suggested by Mankin et al. (2002), a high degree of infestation can be recognized by the pulse threshold of 0.33 pulses/s.

In addition, modern signal-processing techniques (spectrum analyses) and computer rating of infestation likelihood have been developed by different

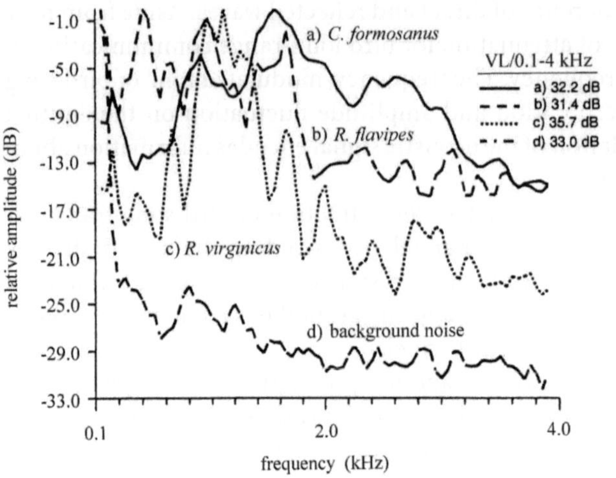

Fig. 7.4. Spectra of signals recorded with an accelerometer for different termite species and urban background noise (Mankin et al. 2002). *a) Coptotermes formosanus, b) Reticulitermes flavipes, c) Reticulitermes virginicus, d)* urban background noise. The vibration level (*VL*) between 0.1 kHz and 4.0 kHz is set on a relative scale

Table 7.1. Termite acoustic emission mean activity rate (pulses/s) measured on standing trees (Mankin et al. 2002)

Species	Symbol	Activity rate at site in quadrant (termite pulses/s)				Mean activity rate at tree	Termite species
		South	East	West	North	(termite pulses/s)	*C. formosanus*
Pine	P3	76.8	90.8	17.0	14.8	49.9	*C. formosanus*
Oak	O1	58.6	30.4	60.9	22.8	43.2	*C. formosanus*
Cypress	C1	0	0	0	74.0	18.5	*C. formosanus*
Oak	O4	0	13.8	30.2	0	11.0	*C. formosanus*
Oak	O2	0	0	40.0	0	10.0	*C. formosanus*
Pine	P1	0	18.7	0	20.1	9.7	*C. formosanus*
Oak	O3	31.5	0	0	0	7.9	*C. formosanus*
Cypress	C2	0	0	21.5	0	5.4	*C. formosanus*
Pine	P4	0	7.5	0	0	1.9	*C. formosanus*
Pine	P2	0	0	0	4.7	1.3	*R. flavipes*
Mean	–	56.0	32.4	34	27.4	–	–

laboratories (Mankin et al. 1996, 2002; Weissling and Thoms 2000) for termite activity detection in situ. The results were completed with measurements in the anechoic chamber. Interesting details on signal recording and signal processing can be seen on the internet at: cmave.usda.us.edu/rmankin/soundlibrary/html.

Difficulties for the acoustic detection of termite activity are related to the signal to noise ratio and the attenuation of signal intensity over distance, as well as the skill of the operator, which is greatly helped by recent technological improvements related to computer signal processing.

7.3 Summary

In the urban forest environment, two major groups coexists: the birds and the insects.

Sound produced by birds in the urban forest environment is perceived in both time and frequency domain. In this environment, the acoustic signals suffer spreading losses when they propagate away from the sender. In the frequency domain, signal propagation leads to an important loss of high energy, due to medium absorption and scatter. When the sender and the receiver are close to the boundary – ground, canopy – the existence of multiple paths modifies the sound propagation parameters. Gradients in air temperature and relative humidity or wind speed cause refraction of the propagating sounds. A sound channel, which enhances propagation, can be created. The overall environmental noise level due to nonhuman sources is between 45 dB and 55 dB SPL.

The most dangerous insects in urban areas are termites. Their detection is a major challenge. Acoustic emission methods operating at more than 40 kHz guarantee their successful detection.

8 Acoustics for Fire Control in Forest

For several years, the number of forest fires all over the world has been increasing. Three main factors are responsible for this phenomenon: long-term drought, the effect of pollution which determines the decline and decay of trees, the formation of loose canopy and the lush growth of grasses which are very inflammable; and another factor is the increasing presence of tourists in forests. For more than 25 years, different remote sensing techniques have been used in forestry (Hildebrandt 1980; Suzuki et al. 2000; Courtier et al. 2001; Angelis et al. 2002).

The remote sensing techniques employ electromagnetic energy (wavelengths from ultraviolet to radio regions) and allow the measurement and analysis of the radiations reflected, transmitted, absorbed and scattered by the atmosphere, by the hydrosphere and by materials on the land surface, for the purpose of understanding and managing the Earth's resources and environment. The first step in the development of remote sensing for the management inventory of forest condition and mapping was oriented to the development of techniques for aerial photos with thermal imagery (e.g. infrared color photos). Radar imagery was also developed, especially for use under difficult meteorological conditions (zone very cloudy or hazy, night observations, etc.). The development of high-resolution imagery obtained with satellites has enormously contributed to the survey of tens of thousands of square kilometers on

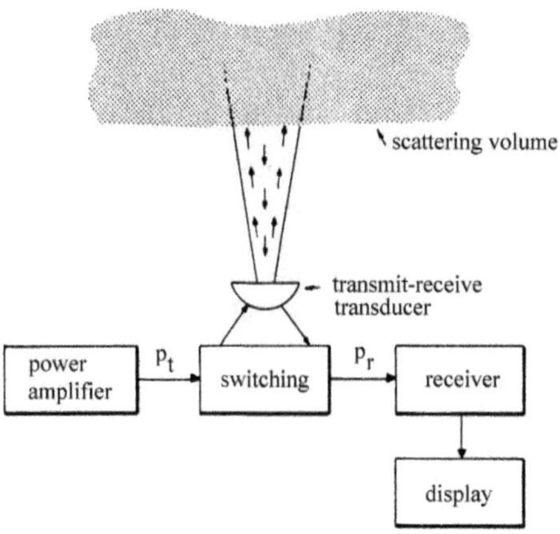

Fig. 8.1. Monostatic acoustic sounder (Beran 1975). Reprinted with permission from the Society of American Foresters, copyright 1975

one image. Regular inspection of the same area with recordings in different wavelengths provides information on the state of forest resources, evaluation of damages, monitoring post-fire forest activities, reforestation and the regeneration of natural forests.

To my knowledge, the first article on acoustic methods for forest fire control was published by the *Journal of Forestry* in 1975 (Beran 1975). The acoustic echo sounder operates as shown in Fig. 8.1, using a single antenna which first transmits a pulse and then switches into receiving mode to collect the

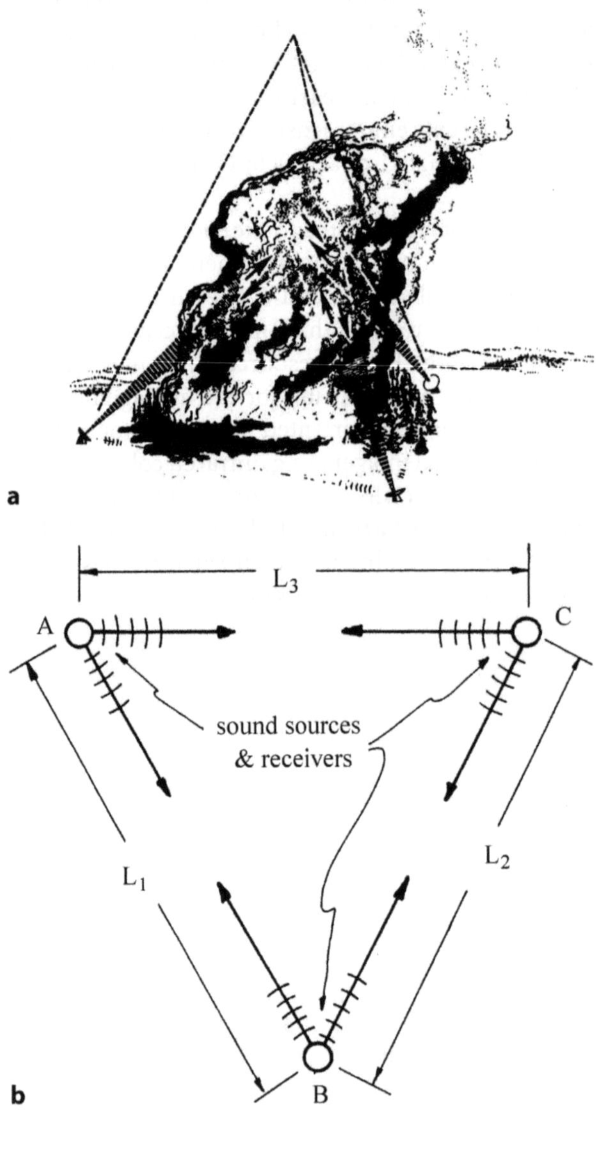

Fig. 8.2. Disposition of receivers for measuring the wind which drives the vortex (Beran 1975). **a** Tridimensional view. **b** Horizontal disposition of the sensors. Reprinted with permission from the Society of American Foresters, copyright 1975

backscattered sound. Variations in the intensity of the signal, directly related to atmospheric turbulence along the path of the original pulse, are recorded as a function of time and indicate the velocity of sound propagating in this field. If three sensors are used, sound scattered from fluctuations in wind turbulence can also be detected. The thermal structure of the boundary layer can be deduced from the monostatic system. The transition from a stable air zone with a horizontal stratification to unstable regions characterized by vertically oriented regions can be identified (Fig. 8.2).

High-quality images and data about the internal structure and external environment of fires and convection columns are necessary for forest fire control. The development of a smoke column in real time, by observing forest fire plume behavior with Doppler lidar (10.6 μm wavelength) and Doppler radar (3.2 cm wavelength), was reported by Banta et al. (1992). The Doppler effect (which is a shift in the acoustic frequency between source and receiver, caused by their relative motion) has been used to detect the kinematics of the convection column and to monitor the smoke plume. The Doppler lidar, which uses a light transmission pulse instead of a radio frequency wave for radar, allows estimation of the radial component of the wind speed. Simultaneous utilization of lidar and radar with two different wavelengths gives information about the size, shape and distribution of particles inside and near the convection column. At the same time, data about topography can be obtained. Tridimensional datasets on kinematics and smoke distribution in the vicinity of the fire helps understanding of the convection column dynamics and smoke plume development.

Satellite data collection and the development of synthetic aperture radars (which are very sensitive to forest biomass), together with the development of multi-frequency microwave sensors for fire and forest monitoring at a local and global scale, are important challenges for the monitoring of forest ecosystems in the near future.

9 Economic Aspects

From previous chapters, we have seen that urban trees are the main components of urban forestry. Indeed, in our days, urban forestry represents a synthesis of policy, planning, landscape architecture and environmental science. Urban trees which add important beauty to our cities and everyday life exist in many towns over the world and have been planted from several centuries ago. As noted in the United States (www.ci.golden-valley.mn.us), "trees are a 'low-tech' solution to energy conservation".

Beside aesthetic considerations, the range of other benefits which come with urban trees, such as environmental, economic and social, is of equal importance. The list of environmental benefits induced by urban and suburban forests is extensive and can include the capacities of trees to remove pollutants from the atmosphere and to contribute to removing greenhouse gasses from the atmosphere. The canopies can act in providing cooling shade, slowing storm water and reducing runoff. The extensive root areas control soil erosion.

Over time, urban trees produce direct and indirect economic benefits. Direct benefits are related to saving energy for cooling homes and increasing property value, as well as reducing airborne emissions from automobiles and other mobile sources. Reducing the air temperature in residential areas in the summer time by several degrees is a well known effect of large shade trees. Wind speed can be reduced, as well as smog. Computer simulation indicates reducing heating costs by 10–30 %.

Indirect benefits are associated with benefits for the commercial sector and retail businesses. A previously cited source noted a rental increase of 7%, an increase of 3–5 % in the sale price of a single family house, an increase in tax revenues and income levels and an increase in the number of jobs and worker productivity. Shady streets are associated with high quality, amenity and comfort; and they enhance economic stability by attracting business and tourists. Treed automobile parking persuades customers to pay a better price. The duration and frequency of visits of these parking places increase. The American Forestry Association (2004) noted that an acre of trees (0.4 ha) can store 2.6 tons of carbon (pollution) annually and generate enough oxygen for 18 people.

Models show (http://envirstudies.brown.edu/classes) that, in 50 years, one tree can generate U.S. $ 30,000 in oxygen, recycle U.S. $ 35,000 in water and remove U.S. $ 60,000 of air pollution. Another indirect benefit is related to stress reduction in the working place and a speed recovery for patients in hospitals.

Helliwell (1967, 1984) was the first to draw attention to the factors which can determine the economics of woodlands and of each stand or tree composing the woodlands. These factors are:

- the quality of the site, its soil and climate, which determine a certain quality of timber produced;
- the value placed on wild life and amenity;
- the quality of present and past management;
- the incidence of taxes.

The woodlands which have existed for a long time (centuries) without significant interruption form the richest habitats for wild plants and animals. Trees provide habitats for birds, mammals, insects, lichens, mosses, etc. As noted by Helliwell (1984) "a mix of oak, ash, lime, maple woodland of fairly open and varied structures is likely to be of high value for nature conservation, and a dense even-aged plantation of exotic evergreen trees is likely to be of relatively low value, but if the evergreen plantation were of more open structure and containing a small proportion of native deciduous trees, its value would probably be much enhanced".

The presence of particular tree species is often less important than the overall structure and management of a woodland. The silvicultural system can even be aged, with clear felling and replanting or with natural seeding, or uneven aged with a two-storied forest – main species or different species in each storey, or with a selection system. The most important factors for the silvicultural systems are:

- the diversity of tree species and type of mixture, including deciduous and evergreen trees;
- the existence of a well developed vertical layer which gives niches for the wild life, together with the presence of old rotten trees that provide habitats for birds, insects, fungi, etc.

To estimate the amenity value of woodlands and trees, Helliwell (1967) proposed a scale going from 1 to 4, which takes into account the following factors: the crown area, the life expectancy, the importance of position in the landscape, the presence of other trees, the form of the species in relation to the setting and the special historical value. This model was refined and enriched by other foresters and scientists, as noted by Bary-Lenger and Nebout (2002), in Europe, Canada, USA, Australia and New Zealand.

The reader interested in detailed calculations of the amenity value of woodlands and trees can refer to: Bary-Lenger and Nebout (2002), Dolwin and Gloss (1993) and Helliwell (1967, 1984).

Arboricultural societies can also be useful, e.g.:

- Europe: the European Arboricultural Council, http://www.eac-arboriculture.com
- USA: http://www2.isa-arbor.com/consumer/value.html
- New Zealand: http://www.rnzih.org.nz/pages/notable.html
- Canada: http://www.Fihoq.qc.ca/html/siaq.html

Table 9.1 lists the benefits produced by trees in urban areas.

Table 9.1. Some benefits produced by trees in urban areas (data from the American Forestry Association, www.ci.golden-valley.mn.us)

Benefit	Beneficial effects of trees	Evaluated costs
Air quality	Leaves absorb carbon dioxide and other atmospheric gases and replenish the air with oxygen for breathing	In 1 year, a mature tree absorbs 26 pounds (11.8 kg) of carbon dioxide and cleans up pollution created by a car driven 11,300 miles (ca. 18 200 km). The same tree also provides enough oxygen for a family of four to breath during an entire year
Water quality	Reduce the impact of rain, diminish soil erosion and runoff into storm sewers. Glare reduction	In 50 years, one tree can recycle U.S.$ 35 000 of water. Reduces storm water run-off by 2%
Energy	Reduce air conditioning and heating needs and increase windbreak protection and shield homes	Reduction of 30% for air conditioning and 25–50% for heating houses
Property economics	Increase property value. Increase the shopping time along tree-lined streets. Apartments and offices in urban wooden areas are rent quicker and have longer leases	Increase property value by 10–20%, increase tax revenues, income level, faster real estate sales, increase the number of jobs
Noise pollution	Absorb sounds, reduce noise intensity	100 foot (30.5 m) width of trees can absorb about 6–8 dB sound intensity
Ecological contribution	Visual screening. Create wildlife diversity. Provide flowers, cones, foliage for decorative and pharmaceutical purposes. Trees are renewable, biodegradable and recyclable. Reduce human stress	Average community tree contribution is U.S.$ 270. Provide habitat for small animals and birds. Increase human productivity in workplaces and offices; and speed recovery in hospitals
Symbolic value	Aesthetics	National heritage

Arboricultural societies can also be useful, e.g.

– Europe: the European Arboricultural Council:
  http://www.eac-arboriculture.com
– USA: http://www2.isa-arbor.com/consumer/value.html
– New Zealand: http://www.nzih.org.nz/pages/notable.html
– Canada: http://www.(?)boq.qc.ca/(?)sing.html

9.4 The ... produced by trees in urban areas

Annex 1 – Symbols

The symbol definitions used in this book are alphabetically arranged, by type of letter, and tabulated below.

Latin Capitals

A, B, C	scales for sound level measurement
D_e	effective trunk diameter
E	Young's modulus [GPa, Pa or N/m^2]
G	coefficient depending on the geometric characteristics of leaves
I	sound intensity [W/m^2]
L	longitudinal or axial growth direction of a tree
L_a	scattering length
L_s	scattering length
FFT	fast Fourier transform
MOE	modulus of elasticity, also E
N	number of trees per square meter
P_s	ambient pressure [Pa]
R	radial direction related to the annual rings
R_e	effective refraction factor
T	absolute temperature [K]
SPL	sound pressure level [dB]
U	voltage
V	railcar speed [km/h]
V_0	the reference speed [60 km/h]
W	watt
Z	mechanical impedance

Script Latin Lower-case Letters

a	half diameter of elliptical cross area of a trunk
b	half diameter of elliptical cross area of a trunk
r_1	length of the direct ray from the source to the receiver
r_2	the reflected ray
c	sound velocity

d	distance
d_0	reference distance
f	frequency
f'	first maximum frequency
f_{pr}	frequency of the associated pressure minimum
i	$\sqrt{-1}$
k	wave number
l	length
m	mass
n	integer
p	sound pressure
$p_A(t)$	instantaneous sound pressure
p_{rms}	root-mean-square sound pressure [Pa or N/m^2]
$p_{reference}$	reference pressure of 20 μPa = 2×10^{-5} N/m^2
r	correlation coefficient
t	time [s]
t_D	growing time of trees [years]
$t_{D_{ref}}$	growing reference-time, trees now aged 12 years
x, y, z	spatial coordinates

Script Latin Capitals

A	excess attenuation
A	acoustic strength
$(Att)_{absorption}$	attenuation produced by leaves absorption
A_d	direct field attenuation
A_s	total scattering attenuation
E_d	energy in the direct field
E_f	energy in the free field
E_r	energy in the reverberant field
$E_{A,T}$	A-weighted sound exposure [Pa^2 s]
E_0	reference A-weighted sound exposure [1.15×10^{-10} Pa^2 s]
F	total surface of leaves per unit volume
I_a	axial inertia momentum
L_A	sound level on A scale [dB]
$L_{E_{A,T}}$	A-weighted noise exposure [dB]
L_A	A-weighted sound pressure [dB]
$L_{d,n}$	day–night noise level [dB]
L_p	sound pressure level; SPL [dB]
L_{eq}	equivalent continuous sound pressure level [dB]
L_{min}	minimum sound pressure level [dB]
L_{max}	maximum sound pressure level [dB]

L_{10}	sound pressure level exceeded 10% of the time [dB]
L_{50}	sound pressure level exceeded 50% of the time [dB]
L_{90}	sound pressure ambient level [dB]
Q_{res}	quality score
S_d	sound level at distance d
S_0	sound level at the source
T	averaging time
$\boldsymbol{T}$	long time over which averaging takes place (h)
I	sound source intensity
I_{ref}	sound source reference intensity $[10^{-12}\ \mathrm{W/m^2}]$
IL or L_{I}	sound intensity level
P_{t}	total pressure at the receiver
P_{d}	direct contribution
P_{r}	speculary reflected contribution
R_{p}	pressure reflection coefficient for a plane wave
W	sound power level [dB $re W_0$]
W_0	the reference sound power $[10^{-12}\ \mathrm{W}]$
Z	ground impedance

Greek Capitals

Φ	diameter
Δt	difference in time
Δ_L	difference in sound pressure level

Greek Lower-case Letters

α_{leaves}	absorption coefficient of leaves
α_m	absorption coefficient of trees
β	relative admittance
λ	wavelength [m]
π	value of pi = 3.14
ν	frequency [Hz]
θ	ambient temperature [°C]
ϱ	density $[\mathrm{kg/m^3}]$
φ	phase angle [°]
σ_{e}	effective flow resistivity $[\mathrm{N\,s\,m^{-4}}]$

Annex 2 – Some Theoretical Considerations

The basic equations and theoretical considerations presented in this section can help the reader in a better understanding of sound propagation in air and are based on my discussion with D. Fellot (personal communication).

In an isotropic medium, the equations of propagation of a plane wave are:

$$\frac{\partial^2 p}{\partial t^2} + c^2 \frac{\partial^2 p}{\partial x^2} = 0 \tag{1}$$

and:

$$\frac{\partial^2 v}{\partial t^2} + c^2 \frac{\partial^2 v}{\partial x^2} = 0 \tag{2}$$

where: c is the propagation velocity, p is the sound pressure, v is the particle velocity, t is time and x is the distance of propagation.

In the air, considered as a perfect gas, in which the variations of atmospheric pressure P and volume V are adiabatic, we have PV^γ = constant and the propagation velocity of an acoustic wave at constant temperature can be calculated with the equation:

$$c^2 = \frac{\gamma P}{\varrho} \quad \text{and} \quad c = \sqrt{\frac{\gamma P}{\varrho}} \tag{3}$$

where γ is the ratio between the specific heat at constant pressure and the specific heat at constant volume and ϱ is air density, 1.293 kg/m^3.

$$\gamma = \frac{C}{c} = 1.4$$

The atmospheric pressure is $P = 1.034 \times 10^5$ Pascal.

From (3), the air velocity has the value $c = \sqrt{1.4 \frac{1.034}{1.293} 10^5} = 334.53$ m/s at a constant temperature of 0 °C.

The sound velocity at a temperature T can be deduced from Marriotte's law, written as:

$$PV = nRT \tag{4}$$

where n is the number of moles, $R = 8{,}314\,\mathrm{J\,kmol^{-1}\,K^{-1}} = 2\,\mathrm{J}$. V, the volume of the air, is:

$$V = \frac{m}{\varrho}\text{, where } m \text{ is the air mass.}$$

Coming back to (4) we have:

$$P\frac{m}{\varrho} = nRT$$

if the temperature is constant, it can be deduced that:

$$\frac{\partial P}{P} - \frac{\partial \varrho}{\varrho} = 0$$

and:

$$2\frac{dc}{c} = \gamma\left(\frac{\partial P}{P} - \frac{\partial \varrho}{\varrho}\right) = 0$$

From previous equations, it can be stated that the sound velocity c is independent of the atmospheric pressure and is strongly dependent on temperature. It can be written as:

$$c^2 = \frac{PV}{m} = \gamma\frac{nRT}{m}$$

the sound velocity c at temperature t [°C] is:

$$c = c_0\sqrt{\left(1 + \frac{t}{273}\right)} \approx c_0\left(1 + \frac{t}{546}\right)$$

where c_0 is the sound velocity at 0 °C.

Because of space limitation, the theoretical aspects treated in this book are limited here.

More theoretical aspects related to the attenuation of sound in air are given by Beranek (1993).

Annex 3 – Frequency Weighting

Frequency weighting is shown in the figure below, which gives the random incidence relative levels as a function of frequency for weighting networks A, B and C (Beranek 1993, p. 808).

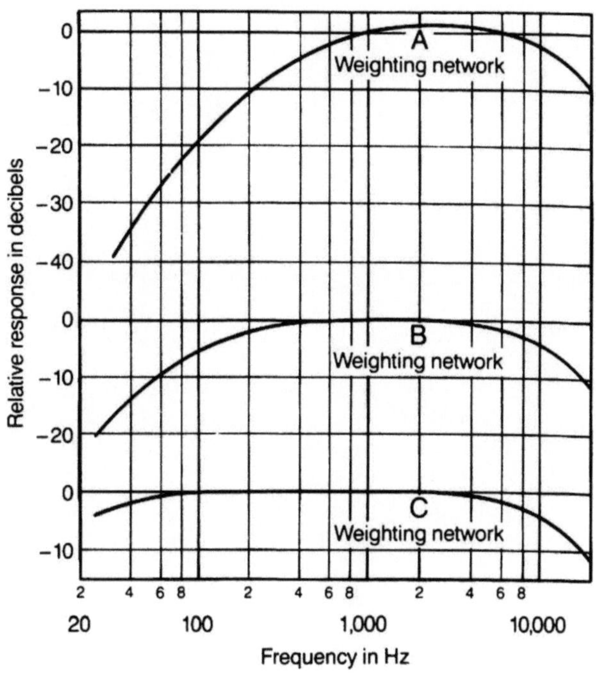

Annex 3 – Frequency Weighting

[illegible]

Annex 4 – Standards

During the past four decades, the introduction of noise control legislation all over the world required the development of appropriate techniques for the measurement of noise produced by outdoor sources.

The reader interest in more details can refer to *Noise Control Engineering Journal* 1987, vol. 29(1), which is entirely devoted to "Measurement Standards", to Harris (1998) and to the International Noise Control Engineering (INCE) collected papers.

The standards listed below are placed in four groups as: (a) acoustics – vocabulary, symbols, units, (b) general noise measurement methods, (c) measurement of specific types of sources, (d) measurement of structures used in noise control.

(a) Acoustics – Vocabulary, Symbols, Units

- ANSI S1. 1-1960 (R 1976) Acoustical terminology
- SAE J 1184 Definition of acoustical terms
- ISO R31/7 Quantities and units of acoustics, 2nd edn, 1992-09-01
- ISO 31-2-1992 Quantities and units. Part 2: periodic and related phenomena, corrected and reprinted 1993-05-15
- ISO/TC 12 quantities, units, symbols, conversion factors
- ANSI/ASME, Y10.11-1084 Letter symbols for acoustics

(b) General Noise Measurement Methods

- ANSI S1. 6-1984 Preferred frequencies, frequency levels and band numbers for acoustical measurements
- ANSI S1. 8-1969 (R1974) Preferred reference quantities for acoustical levels

(c) Measurement of Specific Types of Sources

- ANSI S1. 13-1971 (R1986) Methods for measuring sound pressure levels
- ANSI S1. 34-1980 (R1986) Engineering methods for the determination of sound power levels of noise sources for essentially free-field conditions over a reflecting plane

- ANSI S1. 36-1979 (R1985) Survey methods for the determination of sound power levels of noise sources
- ANSI S12. 37-1988 Determination of sound power levels of sound sources. Methods for in situ measurements using a reference sound source
- ASTM E1014-84 Method for the measurement of outdoor A-weighted sound levels
- SAE J 184-1980 Qualifying a sound data acquisition system
- ISO 2204 Acoustics. Guide to international standards on the measurement of airborne acoustical noise and evaluation of its effect on human beings
- ISO 3740 Acoustics. Determination of sound power levels of noise sources. Guidelines for the use of basic standards and for the preparation of noise test codes
- ISO 3744 Acoustics. Determination of sound power levels of noise sources using sound pressure. Engineering method in an essentially free field over a reflecting plane

(d) Measurement of Structures used in Noise Control

- ANSI S12. 36-1987 Methods for the measurement of acoustical performance of outdoor noise barriers
- ASTM E1014 6 1984 Method for the measurement of outdoor A-weighted sound levels
- ISO 1996-1: 2003 Description measurement and assessment of environmental noise. Part 1 Basic quantities and assessment procedures. Part 2 Acquisition of data pertinent to the land use. Part 3 Application to noise limitations
- ISO 9613 Acoustics. Attenuation of sound during propagation outdoors. Part 1 Calculation of the absorption of sound by the atmosphere. Part 2 General methods of calculation
- IEC 804:1985 integrating averaging sound level meters, Amendment 1: 1989, Amendment 2: 1993

Annex 5 – Units

Commonly used units in this book, presented in alphabetic order:

- bel [B]; 1 B is the level of a power quantity when $\log \frac{P}{P_0} = 1$, P_0 being the reference power. Also 1 B is the level of field quantity when $2 \log \frac{A}{A_0} = 1$; 1 B = 1.151293 Np
- decibel [dB]; 10 dB = 1 B; 1 dB = 0.1151293 Np
- degree Celsius [°C]; for conversion to K, use temperature [°C] = T [K] −273.15
- hertz [Hz]; 1 Hz = s^{-1}
- joule [J]; 1 J = 1 N m = $m^2\,kg\,s^{-2}$
- kilogram [kg]; derived unit kg/m^3
- kelvin [K]
- meter [m]; derived unit m^2, m^3 m/s, m/s^2
- neper [Np]; 1 Np is the level of a field quantities when the logarithm of the ratio of two amplitude is 1 as $\ln \frac{A}{A_0} = 1$
- newton [N]; 1 N = 1 kg m s^{-2}
- octave – the frequency interval between f_1 and f_2, if $\frac{f_1}{f_2} = 2$
- pascal [Pa]; 1 Pa = 1 N m^{-2} = $m^{-1}\,kg\,s^{-2}$
- phone [phone]; 1 phone is the loudness level when $2 \log \frac{p_{\text{eff}}}{p_0} = 0.1$. For a pure tone of frequency 1 kHz, 1 phone ≈ 1 dB
- radian [rad]; 1 rad = mm^{-1}; $1° = \frac{\pi}{180}$ rad, derived unit rad/s
- second [s]
- watt [W]; 1 W = 1 J s^{-1} = $m^2\,kg\,s^{-3}$

The reader interested in measures, units and conversions is invited to visit:

http://www.ex.ac.uk/cimt/dictunit/dictunit.htm (a dictionary of units by F. Tapson)

http://physics.nist.gov./cuu/units (the NIST reference on constants, units and uncertainty)

http://convert-me.com/en (online conversion of weights and units, metric conversions)

References

AFNOR XP S31-133 (2001) Acoustique. Bruit des infrastructures de transports terrestres. Calcul de l'atténuation du son lors de sa propagation en milieu extérieur, incluant les effets météorologiques (French Standard AFNOR XP S31-133). Government Offices, Paris

American Forestry Association (2004) www.ci.golden-valley.mn.us

Anderson GS, Kruze UJ (1992) Outdoor sound propagation. In: Beranek L, Vèr IL (eds) Noise and vibration control engineering. Principles and applications. Wiley, New York, pp 113–144

Angelis CF, Freitas CC, Valeriano DM, Dutra LV (2002) Multitemporal analysis of land use/land cover. JERS-1 backscatter in the Brazilian tropical rain forest. Int J Remote Sensing 23:1231–1240

Attenborough K (1982) Predicted ground effect for highway noise. J Sound Vibr 81:413–424

Attenborough K (1985) Acoustical impedance models for outdoor ground surfaces. J Sound Vibr 99:521–544

Attenborough K (1988) Review of ground effects on outdoor sound propagation from continuous broadband sources. Appl Acoust 24:289–319

Attenborough K (1992) Ground parameter information for propagation modeling. J Acoust Soc Am 92:1042–1046

Attenborough K, Taharzadeh S, Bass HE et al. (1995) Benchmarks cases for outdoor sound propagation models. J Acoust Soc Am 97:173–191

Axmon J, Hansson M (1999) Estimation of circumferential parameters of living trees. In: Namazi NM (ed) Proceedings of the international conference on signal and image processing (SIP'99). SIP, Calgary, pp 188–192

Axmon J, Hansson M (2000) Transients in circular symmetry: estimation of transverse wave propagation velocity. Proc IEEE Sensor Array Multichannel Signal Process Workshop 2000:212–216

Axmon J, Hansson M, Sörnmo L (2004) Experimental study on the possibility of detecting internal decay in standing Picea abies by blind impact response analysis. Forestry 77:179–192

Aylor D (1972a) Noise reduction by vegetation and ground. J Acoust Soc Am 51:197–205

Aylor D (1972b) Sound transmission through vegetation in relation to leaf area density, leaf width and breath of canopy. J Acoust Soc Am 51:411–414

Aylor D (1977) Some physical and psychological aspects of noise attenuation by vegetation. Proceedings of the conference on metropolitan physical environment. Northeastern Forest Experiment Station, USDA Forest Service, Gen Tech Rep NE-25:229–233

Aylor DE, Marks LE (1976) Perception of noise transmitted through barriers. J Acoust Soc Am 59:397–400

Aymonin GG (1986) Guide des arbres et arbustes. Sélection du Reader's Digest, Paris

Banta RM, Olivier LD, Holloway ET, Kropfli RA, Bartram BW, Cupp RE, Post MJ (1992) Smoke-column observations from two forest fires using Doppler Lidar and Doppler Radar. J Appl Meteorol 31:1328–1349

Barrière N, Gabillet Y (1998) Traffic noise control by forests. The meteorological screening effect. Proc Euro-noise 98:1117–1122

Barrière N, Gabillet Y (1999) Sound propagation over a barrier with realistic wind gradient. Acta Acustica 85:325–334

Bary-Lenger A, Nebout JP (2002) Evaluation financière des arbres d'agrément et de production en ville, à la campagne, en forêt. Editions TEC & Doc Lavoisier, Paris

Bass HE, Sutherland LC, Zuckerward AJ (1995) Atmospheric absorption of sound: further developments. J Acoust Soc Am 97:680–683

Beer R (1996) The role of trees in the urban environment: the example of Geneva. Arboricult J 20:437–444

Beran DW (1975) Acoustic echo sounding systems: their potential in forest fire control and research. J For 73:633–639

Beranek L (1960) Noise reduction. McGraw–Hill, New York

Beranek L (1971) Noise and vibration control. McGraw–Hill, New York

Beranek L (1992) Waves and impedances. In: Beranek L, Vèr IL (eds) Noise and vibration control engineering. Principles and applications. Wiley, New York, pp 25–43

Beranek L (1993) Acoustical measurements (revised edn). American Institute of Physics, Melville

Beranek L, Vèr IL (1992) Noise and vibration control engineering. Principles and applications. Wiley, New York

Bies D, Hansen C (1996) Engineering noise control. Theory and application. Spon Press, London

Bies D, Hansen C (2003) Engineering noise control. Theory and practice. Spon Press, London

Bishop DE, Schomer PD (1998) Community noise measurements. In: Harris CM (ed) Handbook of acoustical measurements and noise control, 3rd edn. Acoustical Society of America, Melville, pp 50.1–50.24

Bowlby W (1998) Highway noise prediction and control. In: Harris CM (ed) Handbook of acoustical measurements and noise control, 3rd edn. Acoustical Society of America, Melville, pp 48.1–48.23

Bradbury JW, Vehrencamp SL (1998) Principles of animal communication. Sinauer Associates, Sunderland

Brown EH (1987) Atmospheric acoustics. Encyclopedia of physical science and technology, vol 2. Academic Press, New York, pp 147–163

Brüchert F, Becker G, Speck T (2000) The mechanics of Norway spruce. Mechanical properties of standing trees from different thinning regimes. For Ecol Manage 135:45–62

Bucur V (1985) Ultrasonic hardness and X-ray densitometric analysis of wood. Ultrasonics 23:269–275

Bucur V (1995) Acoustics of wood. CRC Press, Boca Raton

Bucur V (2003) Nondestructive characterization and imaging of wood. Springer, Berlin Heidelberg New York

Bucur V (2005) Acoustics of wood, 2nd edn. Springer, Berlin Heidelberg New York

Bullen R, Fricke F (1982) Sound propagation through vegetation. J Sound Vibr 80:11–23

Burns SH (1979) The absorption of sound by pine trees. J Acoust Soc Am 65:658–661

Canévet G (1996) La psychoacoustique. Aspects fondamentaux et examples d'applications. Acoust Techn 5:7–14

Carlson DE, McDaniel OH, Reethof G (1977) Noise control by forests. Proc Int Noise 77:B576–B586

Chappuis C (1971) Un exemple de l'influence du milieu sur l'émission vocale des oiseaux. L'évolution des chants en forêt équatoriale. Terre Vie 25:183–202

Chuang ST, Wang SY (2001) Evaluation of standing tree quality of Japanese cedar grown with different spacing using stress-wave and ultrasonic wave methods. J Wood Sci 47:245–253

Cole L, Mullard J (1982) Woodlands in urban areas – a resource and refuge. Arboricult J 6:295–300

Cook DI, Haverbeke DF van (1971) Trees and shrubs for noise abatement. (Research Bulletin 246.) Agricultural Experiment Station, University of Nebraska, Nebraska City, 77 pp

Cook DI, Haverbeke DF van (1972) Trees, shrubs and land-forms for noise control. J Soil Water Conserv 27:259–261

Cook DI, Haverbeke DF van (1977) Suburban noise control with plant materials and solid barriers. Northeastern Forest Experiment Station, USDA Forest Service, Gen Tech Rep NE-25:234–241

Council of Europe (2004) European urban charter, chap 4.2. Environment and nature in town. http://www.coe.int/T/E/Clrae_5._Texts/5._Convention and charters/urban.aspects (visited March 2005)

Courtier S, Taylor D, Siegert F, Hoffmann A, Bao MQ (2001) ERS SAR backscatter. A potential real-time indicator of the proneness of modified rainforests to fire. Remote Sensing Environ 76:410–417

Crombie DH, Hothersall DC, Chandler-Wilde SN (1995) Multiple edge noise barriers. Appl Acoust 44:353–387

Decourt N (1975) The attenuation of noise by vegetation. Rev For Fr 27:419–429

Defrance J, Barrière N, Premat E (2002) Forest as a meteorological screen for trafic noise. Proceedings of the 9th international congress on sound and vibration. Orlando, CD-Rom

DeJong R, Stusnick E (1976) Noise control studies of the effect of wind on acoustic barrier performance. Noise Control Eng 1976:101–109

Dolwin JA, Gloss CL (1993) Evaluation of amenity trees mainly in private ownership within the borough of Tunbridge Wells. Arboricult J 17:301–308

Duhamel D, Sergent P, Hua C, Cintra D (1998) Measurement of active control efficiency around noise barriers. Appl Acoustics 55:217–241

Egan D (1988) Architectural acoustics. McGraw–Hill, New York

Embelton TFW (1963) Sound propagation in homogeneous deciduous evergreen woods. J Acoust Soc Am 35:1119–1125

Embelton TFW (1966) Scattering by an array of cylinders as a function of surface impedance. J Acoust Soc Am 40:667–670

Embelton TFW (1996) Tutorial on sound propagation outdoors. J Acoust Soc Am 110:31–48

EN 1793-3(1997) Road traffic noise reducing devices – test method for determining the acoustic performance, part 3: normalized traffic noise spectrum (European Standard EN 1793-3). European Commission, Brussels

European Commission (1996) Future noise policy (Green paper, Directorate general for environment, nuclear safety and civil protection). European Commission, Brussels

Eyring CF (1946) Jungle acoustics. J Acoust Soc Am 18:257–270

Fahy FJ, Walker JG (1998) Fundamentals of noise and vibration. E & FN Spon, London

Fang CF, Ling DL (2003) Investigation of the noise reduction provided by tree belts. Landscape Urban Planning 63:187–195

Fricke F (1984) Sound attenuation in forests. J Sound Vibr 92:149–158

Fyhri A, Klaeboe A (1999) Exploring the visual aesthetics on the soundscape. Proc Int Noise Fort Landerdale 2:1261–1264

Gautier PE, Poisson F, Foy F (2004) Les enjeux actuels des recherches en acoustique et vibrations dans le domaine ferroviaire. Acoust Techn 37:6–11

Ge J, Hokao K (2004) Research on the sound environment of urban open space from the viewpoint of soundscape – a case study of Saga Forest Park, Japan. Acta Acustica 90:555–563

Goodfriend LS (1977) Range of sound levels in the outdoor environment. Proc Conf "Metropolitan physical environment" Syracuse, New York, pp 172–182

Gygi B, Kidd GR, Watson CS (2004) Spectral temporal factors in the identification of environmental sounds. J Acoust Soc Am 115:1252–1265
Haglof (2003) Vertex laser instrument; www.haglofsweden.com
Hanson CE, Saurenman HJ, Towers DA (1998) Rail transportation and vibration. In: Harris CM (ed) Handbook of acoustical measurements and noise control, 3rd edn. Acoustical Society of America, Melville, pp 46.1–46.24
Harris CM (1998) Handbook of acoustical measurements and noise control, 3rd edn. Acoustical Society of America, Melville
Haupt R (1973) Noise reduction by forest stands, part I: fundamentals of the method of investigation. Arch Naturschutz Landschaftsforsch 13:309–327
Haupt R (1974) Noise reduction by forest stands, part 2: noise reduction by forest stands. Arch Naturschutz Landschaftsforsch II 14:61–75
Heimann D (2003) Numerical simulation of wind and sound propagation through an idealised stand of trees. Acta Acustica 89:779–788
Heisler GM (1974) Trees and human comfort in urban areas. J For 72:466–469
Helliwell DR (1967) The amenity value of trees and woodlands. Scot For 21:109–112
Helliwell DR (1984) Economics of woodland management. Packard Publishing, Chichester
Herrington LP (1974) Trees and acoustics in urban area. J For 72:462–465
Herrington LP, Brock C (1977) Propagation of noise over and through a forest stand. Proceedings of the conference on metropolitan physical environment. Northeastern Forest Experiment Station, USDA Forest Service, Gen Tech Rep NE-25:226–228
Heutschi K (2004) SonRoad. New Swiss road traffic noise model. Acta Acustica 90:548–554
Hildebrandt G (1980) Survey of remote sensing techniques in forestry. In: Fraysee G (ed) Remote sensing application in agriculture and hydrology. Commission of the European Community, Ispra, pp 25–36
Hora B (1981) The Oxford encyclopaedia of trees of the world. Oxford University Press, Oxford
Hosie RC (1969) Native trees of Canada. Canadian Forestry Service, Ottawa
Huang CL, Lindström H, Nakada R, Ralston J (2003) Cell wall structure and wood properties determined by acoustics – a selective review. Holz Roh Werkst 61:321–325
Huddart L (1990) The use of vegetation for traffic noise screening. (Transport Research Laboratory Report RR 238.) Transport Research Laboratory, Crowthorne
Huisman WHT (1989) Reverberation and attenuation by trees: measured and modeled. Paper at Spring Conference, Institute of Acoustics, Oxford, pp 6
Huisman WHT (1990) Sound propagation over vegetation – covered ground. PhD thesis, Katholieke Universiteit van Nijmengen, Nijmengen
Huisman WHT, Attenborough K (1991) Reverberation and attenuation in pine forest. J Acoust Soc Am 90:2664–2677
Huisman WHT, Martens MJM, Beckwilder RWB, Londers AMJA (1988) Measured and modeled sound transmission in a pine forest. Proc Int Noise 88:1583–1586
Ingård V (1953) A review of the influence of meteorological conditions on sound propagation. J Acoust Soc Am 25:405–411
Ishii M (1994) Measurement of road traffic noise reduced by the employment of low physical barriers and potted vegetation. Int Noise 29:595–597
Ishizuka T, Fujiwara K (2004) Performance of noise barriers with various edge shapes and acoustical conditions. Appl Acoust 65:125–141
ISO 9613-2 (1996) Acoustics – attenuation of sound during propagation outdoors, part 2: general method of calculation (ISO 9613-2). ISO, Geneva
Jacques D, Marchal M, Curnel Y (2004) Relative efficiency of alternative methods to evaluate wood stiffness in the frame of hybrid larch (*Larix eurolepsis*, Henry) clonal selection. Ann For Sci 61:35–43

Jimenez Altamirano F (1997) Acoustic plant screens as a buffer against traffic noise. Arquit Paisaje 40:20–21

Kellomäki S, Haapanen A, Salonen H (1976) Tree stands in urban noise abatement. Silva Fennica 10:237–256

Kinsler LE, Frey AR, Coppens AB, Sanders JV (1982) Fundamentals of acoustics. Wiley, New York

Koizumi A (1987) Studies on the estimation of the mechanical properties of standing trees by non-destructive bending test. Bull Coll Exp For Fac Agric Hokkaido Univ 44:1329–1415

Koizumi A (1990) Variability in wood quality of Japanese larch observed by tree bending tests. Proc IUFRO Congr 19:1–7

Kotzen B (2004) Plants and environmental noise barriers. Acta Hort 643:265–273

Kragh J (1979) Pilot study on railway noise attenuation by belt of trees. J Sound Vibr 66:407–415

Kragh J (1981) Road traffic noise attenuation by belt of trees. J Sound Vibr 74:235–241

Krog NH, Engdahl B (2004) Annoyance with aircraft noise in local recreational areas, contingent on changes in exposure and other context variables. J Acoust Soc Am 116:323–333

Kroodsma DE, Miller EH, Ouellet H (1982) Acoustic communication in birds. Academic Press, New York

Kuc R (2001) Transforming echoes into pseudo-action potentials for classifying plants. J Acoust Soc Am 110:2198–2206

Kuttruff H (1967) Concerning reverberation in media with randomly distributed scattering centers, particularly in reverberation rooms with suspended scattering elements. Acustica 18:131–143

Lanbourg P (1989) Rigidité de l'arbre sur pied, indicateur de l'élasticité longitudinale du bois, application aux peupliers. PhD thesis, Inst Nat Polytechnique de Lorraine, Nancy

Launay J, Rozenberg P, Pâques L et al. (2000) A new experimental device for rapid measurement of the trunk equivalent modulus of elasticity on standing trees. Ann For Sci 57:351–359

Laverty R (2001) Mechanics of layered cylindrical elastic waveguides. PhD thesis, Colorado State University, Denver

Leininger TD, Schmoldt DL, Tainter FH (2001) Using ultrasound to detect defects in trees: current knowledge and future needs. Proc Int Precision For Coop Symp 1:99–107

Leonard RE, Herrington LP (1971) Noise abatement in a pine plantation. Northeastern Forest Experiment Station, USDA Forest Service, Res Note NE-140:1–6

Lindström H, Harris P, Nakada R (2002) Method for measuring stiffness of young trees. Holz Roh Werkst 60:165–174

Lyon RH, Blair CN, DeJong RG (1977) Evaluating effects of vegetation on the acoustical environment by physical scale-modeling. Proceedings of the conference on metropolitan physical environment. Northeastern Forest Experiment Station, USDA Forest Service, Gen Tech Rep NE-25:218–225

Mace BL, Bell PA, Loomis RJ (1999) Aesthetic, affective and cognitive effects of noise on natural landscape assessment. Soc Nat Resour 12:225–242

Mamdy C, Rozenberg P, Franc A et al. (1999) Genetic control of stiffness of standing Douglas-fir. Ann For Sci 56:133–144

Mankin RW, Shuman D, Coffelt JA (1996) Noise shielding of acoustic devices for insect detection. J Econ Entomol 89:1301–1308

Mankin RW, Osbrink WL, Oi, FM, Anderson JB (2002) Acoustic detection of termites infestation in urban trees. J Econ Entomol 95:981–988

Mao LS, Gao Y, Sun WQ (1993) Influences of street tree systems on summer micro-climate and noise attenuation in Nanjing City, China. Arboricult J 17:239–251

Marquis-Favre C, Premat E, Aubrée D (2005a) Noise and its effects – a review on quantitatrive aspects of sound, part 1: notations and acoustic rating. Acta Acustica 91:613–625

Marquis-Favre C, Premat E, Aubrée D (2005b) Noise and its effects – a review on quantitatrive aspects of sound, part 2: noise and annoyance. Acta Acustica 91:626–642

Marten K, Marler P (1977) Sound transmission and its significance for animal vocalization I. Temperate habitats. Behav Ecol Sociobiol 2:271–290

Marten K, Quine D, Marler P (1977) Sound transmission and its significance for animal vocalization. II. Tropical forest habitats. Behav Ecol Sociobiol 2:291–302

Martens MJM (1977) The influence of the soil on the acoustic climate of plant communities. Proc Int Noise 77:B593–B598

Martens MJM (1980) Foliage as a low-pass filter:experiments with model forests in anechoic chamber. J Acoust Soc Am 67:66–72

Martens MJM (1981) Noise abatement in plant monocultures and plant communities. Appl Acoust 14:167–189

Martens MJM, Michelsen A (1981) Absorption of acoustic energy by plants leaves. J Acoust Soc Am 69:303–306

Martens MJM, Heijden LAM van der, Walthaus HHJ, Rens WJJM van (1985) Classification of soils based on acoustical impedance, air flow resistivity and other physical soil parameters. J Acoust Soc Am 78:970–980

Martinis R (2002) Analisi e svilupo di techniche non invasive per valutazione di carie in alberi in piedi. PhD thesis, University of Florence, Florence

Martinis R, Socco V, Sambuelli L, Nicolotti G, Schmitt O, Bucur V (2004) Tomographie ultrasonore pour les arbres sur pied. Ann For Sci 61:157–162

Mattheck C (1996) Tree – the mechanical design. Springer, Berlin Heidelberg New York

Mattheck C (1998) Design in nature – learning from trees. Springer, Berlin Heidelberg New York

Mattheck CG, Bethge KA (1993) Detection of decay in trees with the Metriguard stress wave timer. J Arbor 19:374–378

McCornnell RL, Eelliott KC, Blizzard SHH, Koster KH (1984) Electronic measurement of tree-row volume. Proc Nat Conf Field Equip Irrig Drain 1:85–90

McKerrow PJ, Harper NL (1999) Recognizing leafy plants in air-sonar. Sensor Rev 19:202–206

Mecklenberg RA, Rintelman WF, Schumaier DR DenBrink CV, Flores L (1972) The effect of plants on microclimate and noise reduction in the urban environment. Hort Sci 95:61–64

Morfey CL (2001) Dictionary of acoustics. Academic Press, San Diego

Morton ES (1975) Ecological sources of selection on avian sounds. Am Nat 108:17–34

Mueller R, Kuc R (2000) Foliage echoes:a probe into the ecological acoustics of bat echolocation. J Acoust Soc Am 108:836–845

Muir J (1984) Silvicultural information to help select and manage native trees in urban or suburban developments. Arboricult J 8:13–17

Mulligan BE, Lewis SA, Faupel ML, Goodman LS, Anderson LM (1987) Enhancement and masking of loudness by environmental factors. Vegetation and noise. Environ Behav 19:411–443

Muradali A, Fyfe KR (1999) Accurate barrier modelling in the presence of atmospheric effects. Appl Acoust 56:157–182

Nadezhdina N, Cermak J (2003) Instrumental methods for studies of structure and function of root systems of large trees. J Exp Bot 54:1511–1521

Nakamura N (1996) Measurement of the properties of standing trees with ultrasonics and mapping of the properties. Bull Tokyo Univ For 96:125–135

Naz P, Parmentier G, Canard-Caruana S (1992) Propagation acoustique à grande distance dans les basses couches de l'atmosphère. J Phys Colloq 2:745

Noble JM, Bass HE, Raspet R (1992) The effect of large scale atmospheric inhomogeneities on acoustic propagation. J Acoust Soc Am 92:1040–1046

Ohsato S (1972) Masking of a 2000 Hz pure tone by wind noise in forest (1). Bull Utsunomiya Univ For 9:27–32

Ouïs D (2001) Detection of rot on standing trees by means of an acoustic technique. Arboricult J 25:117–152

Padgham M (2004) Reverberation and frequency attenuation in forests – implications for acoustic communication in animals. J Acoust Soc Am 115:402–410

Pal AK, Kumar V, Saxena NC (2000) Noise attenuation by green belts. J Sound Vibr 234:149–165

Papafotiou M, Chronopoulos J, Tsiotsios, Mouzakis K, Balotis G (2004) The impact of traffic noise control in an urban park. Acta Hort 643:277–279

Pellerin RF (1965) A vibrational approach to NDT of structural lumber. For Prod J 15:93–101

Pellerin RF, Ross RJ (2002) Nondestructive evaluation of wood. Forest Products Society, Madison

Perfater MA (1979) Community perception of noise barriers, vol 1. Virginia Highway and Transportation Research Council, Richmond

Piccolo A, Plutino D, Cannistraro G (2005) Evaluation and analysis of the environmental noise of Messina, Italy. Appl Acoust 66:447–465

Piercy JE, Daigle GA (1998) Sound propagation in the open air. In: Harris CM (ed) Acoustical measurements and noise control, 3rd edn. Acoustical Society of America, Melville, pp 3.1–3.26

Piercy JE, Embelton TFW, Sutherland LC (1977) Review of noise propagation in atmosphere. J Acoust Soc Am 61:1403–1418

Piersol D (1992) Data analysis. In: Beranek LL, Vér IL (eds) Noise and vibration control engineering. Principles and applications. Wiley, New York, pp 45–74

Price MA (1988) Sound attenuation in woodland. PhD thesis, Open University, Milton Keynes

Price MA, Attenborough K, Heap NW (1988) Sound attenuation through trees: measurements and models. J Acoust Soc Am 84:1836–1844

Raimbault M, Lavandier C, Cremezi-Chharlet C, Walker E (2004) Situations de multi-exposition aux bruits routier et ferroviaire: quelles pistes de recherche pour de nouveaux indicateurs? Acoust Tech 37:22–27

Raney JP, Cawthorn JM (1998) Aircraft noise. In: Harris CM (ed) Handbook of acoustical measurements and noise control, 3rd edn. Acoustical Society of America, Melville, pp 47.1–47.21

Raynor GS (1971) Wind and temperature in a coniferous forest and contiguous field. For Sci 17:351–363

Reethof G, Heisler GM (1976) Trees and forest for noise abatement and visual screening. Northeastern Forest Experiment Station, USDA Forest Service Gen Tech Rep NE-22:39–48

Reethof G, McDaniel OH, Heisler GM (1977) Sound absorption characteristics of tree bark and forest floor. Proceedings of the conference on metropolitan physical environment. Northeastern Forest Experiment Station, USDA Forest Service, Gen Tech Rep NE-25:206–216

Reynolds KD, Wilson JN (1989) Ultrasonics for height measurements on forestry equipment. Am Soc Agric Eng Pap 89:7541

Rizzo DM, Gross R (2000) Distribution of *Armillaria melea* on pear roots systems and comparison of excavation techniques. In: Stokes A (ed) The supporting roots of trees and woody plants: form, function and physiology. (Development in plant and soil sciences, vol 87.) Kluwer Academic, Dordrecht, pp 305–311

Rogers JC, Lee SM (1989) Sound scattering from forests:considerations and initial measurements. Proc Int Conf Acoust 14:277–281

Rogers JC, Shen W, Lee SM (1990) Acoustic scattering:measurements on model trees and their extension to natural forests. Proc Inst Acoust 12:827–833

Rogers JC, Chen P, Mueller AW (1992) A study of impulse source scattering in a forest. Proceedings of the 14th international conference on acoustics, abstract, Beijing. Acustica 76:87

Ross RJ (1985) Stress wave propagation in wood products. Proceedings of the 5th symposium on nondestructive testing on wood. Washington State University, Pullman, pp 291–318

Ross RJ, Pellerin RF (1991) Nondestructive evaluation of wood:past, present and future. In: Ruud CO, Green RE (eds) Nondestructive characterization of materials, vol 4. Plenum Press, New York, pp 59–64

Ross RJ, Pellerin RF (1994) Nondestructive testing for assessing wood members in structures. A review. USDA Forest Product Laboratory, Madison, Gen Tech Rep FPL-GTR-70

Salomons EM (2001) Computational atmospheric acoustics. Kluwer, Doordrecht

Schaudinischky LH; Schiller G; Keller Y (1982) Acoustic properties of individual trees and forest strips. Forstarchiv 53:52–57

Scholes WE, Salvidge AC, Sargent JW (1971) Field performance of a noise barrier. J Sound Vibr 16:627–642

Schultz TJ (1978) Synthesis of social surveys on noise annoyance. J Acoust Soc Am 64:377–405

Slabbekoorn (2004) Habitat-dependent ambient noise: consistent spectral profiles in two African forest types. J Acoust Soc Am 116:3727–3733

Steele C (2001) A critical review of some noise prediction model. Appl Acoust 62:271–287

Stephens RWB, Bate AZ (1966) Acoustics and vibrational physics. Arnold, London

Stokes A, Fourcaud T, Hruska J, Cermak J, Nadezhdina N, Nadezhdin V, Praus L (2002) An evaluation of different methods to investigate root system architecture of urban trees in situ 1. Ground-penetrating radar. J Arboricult 28:2–10

Suzuki Y, Kimura K, Aoki T (2000) A GIS based study toward forecast of suburban forest change. Geol J 52:203–212

Takata K, Teraoka Y (2002) Genotypic effects on the variation of wood quality and growth traits in plantation forest made by cutting cultivars of Japanese cedar. J Wood Sci 48:106–113

Tanaka K, Ikeda S, Kimura R, Simazawa K (1979) The function of forests in soundproofing. Bull Tottori Univ For 11:77–102

Upchurch BL, Anger WC, Vass G, Glenn DM (1992) Ultrasonic tree caliper. Appl Eng Agricult 8:704–714

Upchurch BL, Glenn DM, Vass G, Anger WA (1993) An ultrasonic tree trunk diameter caliper. Hortic Tech 3:89–91

USDA Agroforestry Center (2004) Trees as noise buffers. (The overstory 60, Agroforestry.net) www. agroforestryNet/oversr (visited in 2005)

Van Kamp I, Job RFS, Hatfield J, Haines M, et al (2004) The role of noise sensitivity in the noise-response relation: a comparison of three international airport studies. J Acoust Soc Am 116:3471–3479

Van Renterghem T, Botteldooren D (2002) Effect of a row of trees behind noise barriers in wind. Acta Acustica 88:869–878

Van Renterghem T, Botteldooren D (2003) Numerical simulation of the effect of trees on downwind noise barrier performance. Acta Acustica 89:764–778

Van Renterghem T, Botteldooren D, Cornelis W, Gabriels D (2002) Reducing screen-induced refraction of noise barriers in wind by vegetative screens. Acta Acustica 88:231–238

Von Gierke HE, Eldred KM, Break RK (1998) Environmental impact statements. In: Harris CM (ed) Handbook of acoustical measurements and noise control, 3rd edn. Acoustical Society of America, Melville, pp 54.1–54.22

Von Wendorff GB (1974) Schützt der Wald vor Fluglärm. (Protection against aircraft noise with forests.) Forstarchiv 45:6–11

Wang X, Ross RJ (2002) Nondestructive evaluation of green materials – recent research and development activities. In: Pellerin RF, Ross RJ (eds) Nondestructive evaluation of wood. Forest Products Society, Madison, pp 149–171

Wang X, Ross RJ, McCellan M, Barbour RJ, Erickson JR, Forsman JW, McGinnis CD (2000) Strength and stiffness assessment of standing trees using a nondestructive stresswave technique. USDA Forest Service, Forest Products Laboratory, Madison, Res Rep FPL-RP-585

Watanabe T, Yamada S (1996) Sound attenuation through absorption by vegetation. J Acoust Soc Jpn 17:175–182

Watts G, Crombie DH, Hothersall DC (1994) Acoustic performance of new designs of traffic noise barriers – full-scale tests. J Sound Vibr 177:289–305

Watts G, Chinn L, Godfrey N (1999) The effects of vegetation on the perception of traffic noise. Appl Acoust 56:39–56

Waxler R (2004) Modal expansion for sound propagation in the nocturnal boundary layers. J Acoust Soc Am 115:1437–1448

Weissling TJ, Thoms EM (2000) Use of an acoustic emission detector for locating Formosan subterranean termite feeding activity when installing and inspecting aboveground termite bait stations. Fla Entomol 82:60–71

Wempen J (1986) Outdoor sound propagation close to the ground. In: Martens MJM (ed) Sound propagation in forested areas and shelter belts. Workshop Proceedings, Nijmegere, pp 83–92

Wempen J, Mellert K (1990) Messung der Impedanz eines Erdbodens mit dem Reflexionsfaktor für ebene Wellen und für Kugelwellen. Acustica 71:1–11

Wiley RH, Richards DG (1978) Physical constraints on acoustic communication in the atmosphere: implications for the evolution of animal vocalization. Behav Ecol Sociobiol 3:69–94

Wiley RH, Richards DG (1982) Adaptations for acoustic communication in birds: sound transmission and signal detection. In: Kroodsma DE, Miller EH, Ouellet H (eds) Acoustic communication in birds, vol 1. Academic Press, New York, pp 131–181

Wilson BF (1984) The growing tree. University Press, Amherst, Mass.

Yamada S, Watanabe T, Nakamura S, Yokoyama H, Takeoka S (1977) Noise reduction with vegetation. Int Noise 77:B599–B606

Zimmermann MH, Brown CL (1980) Trees – structure and function. Springer, Berlin Heidelberg New York

Zipper WC, Rowntree RA, Stevens JC (1991) Structure and composition of street-side trees of residential areas in state of Maryland, USA. Arboricult J 15:1–11

Subject Index

MIX
Papier aus verantwortungsvollen Quellen
Paper from responsible sources
FSC® C105338

If you have any concerns about our products,
you can contact us on
ProductSafety@springernature.com

In case Publisher is established outside the EU,
the EU authorized representative is:
Springer Nature Customer Service Center GmbH
Europaplatz 3, 69115 Heidelberg, Germany

Printed by Libri Plureos GmbH
in Hamburg, Germany